Amelia Island

Birthplace of the Modern Shrimping Industry

Gray Edenfield

FONTHILL

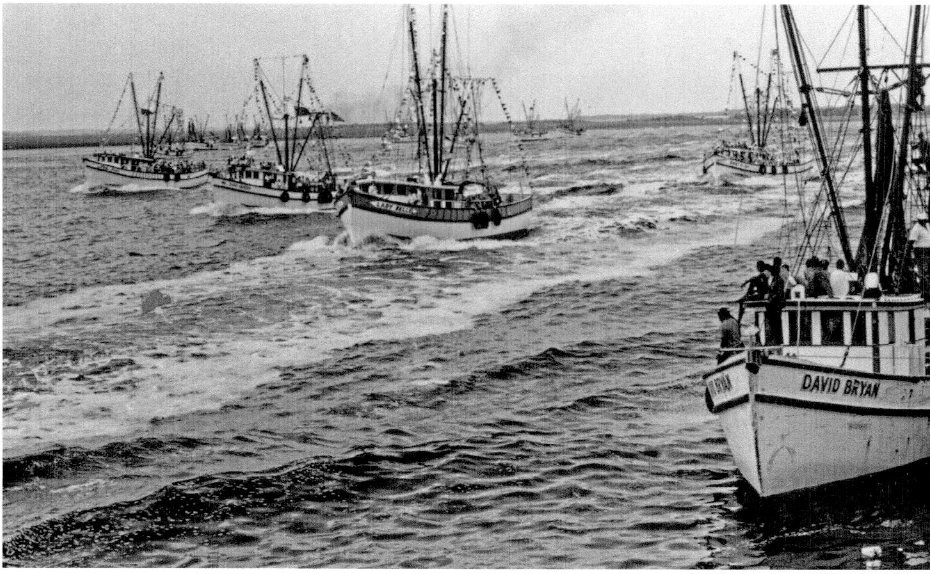

The *Chickadee* captained by Sloan Peterson, on the way to a 3rd place finish in 1969.
AIMH Archives

Dedicated to the memory of Barbara Alexander Gray – the tough and truly special southern lady who made Fernandina my birthright

Fonthill Media LLC
www.fonthillmedia.com
office@fonthillmedia.com

First published 2015

ISBN 978-1-63499-005-9

Typeset in 10pt on 13.5pt Minion Pro
Printed and bound in England

Contents

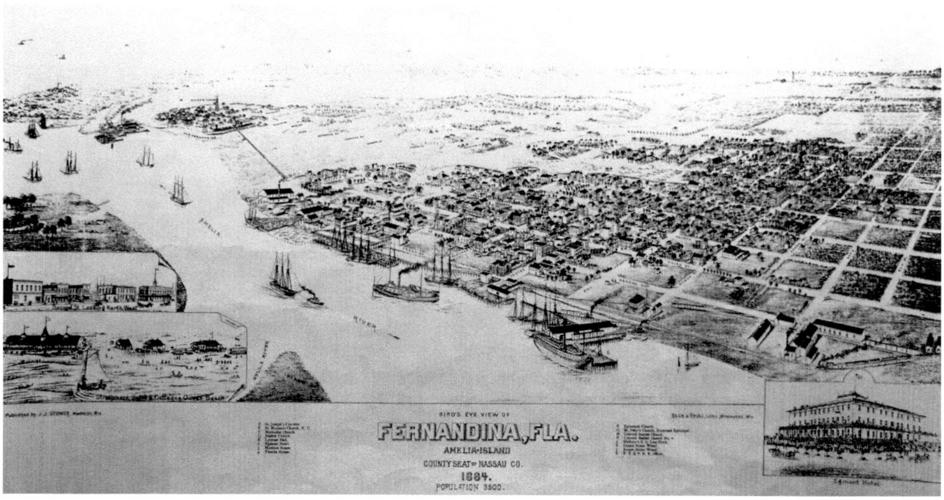

Sketch of Downtown Fernandina in 1884. Amelia Island Museum of History Archives.

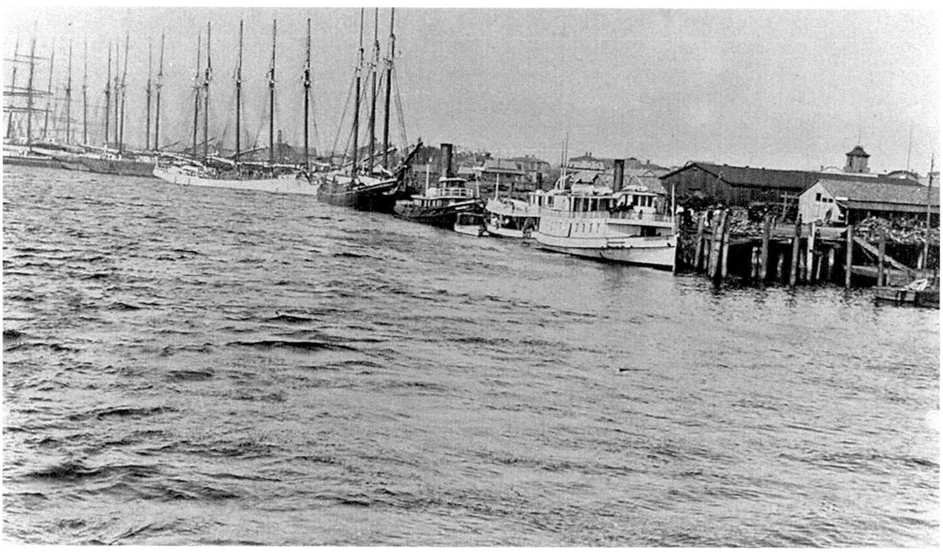

Downtown Fernandina's harbor in the late Victorian Era, with a side view of the yacht *Skibo*, which belonged to the family of Scottish-born industrialist Thomas Carnegie. The Carnegies owned much of the land on nearby Cumberland Island. *Amelia Island Museum of History Archives*

Introduction

Shrimp have been caught around the world for thousands of years, and have always been plentiful off the coast of Amelia Island. Before the 1900s shrimp were caught close to shore, using nets (called haul seines) hung between rowboats. In 1879-80 the United States Government, in its first fishery statistics report said of Fernandina: "shrimp and prawn are abundant in the harbor directly opposite the city during the entire year, and a man can readily secure three or four bushels with a small cast-net on any pleasant night. The catch, which is not less than forty bushels is boiled and dried for shipment to New York, Philadelphia, and Savannah."

However, the amount of shrimp being caught on Florida's East Coast paled in comparison to the Gulf Coast's catch. In 1880 the entire East Coast produced 72,000 pounds of shrimp valued at $4,000, while the state of Louisiana alone produced 534,000 pounds, valued at $42,000. While Florida would always trail the Bayou State in shrimp production, Florida's northernmost barrier island would make contributions to shrimping technology matched by no other place in the world.

At the beginning of the twentieth century, several innovations made by notable Fernandina residents revolutionized the technology of shrimping, and made Fernandina the "Birthplace of the Modern Shrimping Industry." The story of the shrimping industry in Fernandina is a multicultural one. Local families who had been on Amelia Island for generations like the Joneses, Traeyes, Cooks, and Littles were joined by Italians named Litrico, Gianino, Serra, and Basetta; Scandinavians like Johnson, Janson, Hansen, and Olsen; and the Portuguese Carinhas brothers along with countless others who followed the shrimp to Amelia Island. This influx of culture and ideas resulted in a diverse and unique community, flush with pioneers and entrepreneurs. These people were the American Dream personified, an assortment of bold, intrepid people from disparate backgrounds who came here with very little, and through ingenuity or necessity carved out a life for themselves on the water.

THE LOCAL SHRIMPING INDUSTRY PRIOR TO 1900

Before the birth of the modern shrimping industry, the biggest name to emerge in local shrimping was T. E. Fisher. Fisher was a Fernandina resident and seafood dealer who established a trade network to markets of New York City around 1875. Outside of Fisher's operation, most of the shrimp caught in the area at that time was sold locally. The market for shrimp was nowhere near what it would later become, but in New York – atypical for the times – a pound of shrimp could be sold for fifty cents.

There were about a dozen shrimpers in the area, who operated rowboats in teams of two. One man would row the boat, as the other operated a cast net, usually ten to fifteen feet in diameter, with half-inch mesh. They shrimped in two main seasons, one beginning in March and lasting about a month, and the other spanning September through December. Most shrimping was done at night, and a usual haul could be up to 136 pounds. The primitive method of preservation at that time was to boil the shrimp in brine and then dry them in the sun on a wooden platform before shipping.[1]

Fisher built up his distribution network to include Philadelphia, Savannah, Atlanta, Macon, and Charleston. By 1879 he was shipping out 29,000 pounds of shrimp annually. The next year he abruptly and somewhat mysteriously ended the venture. When interviewed

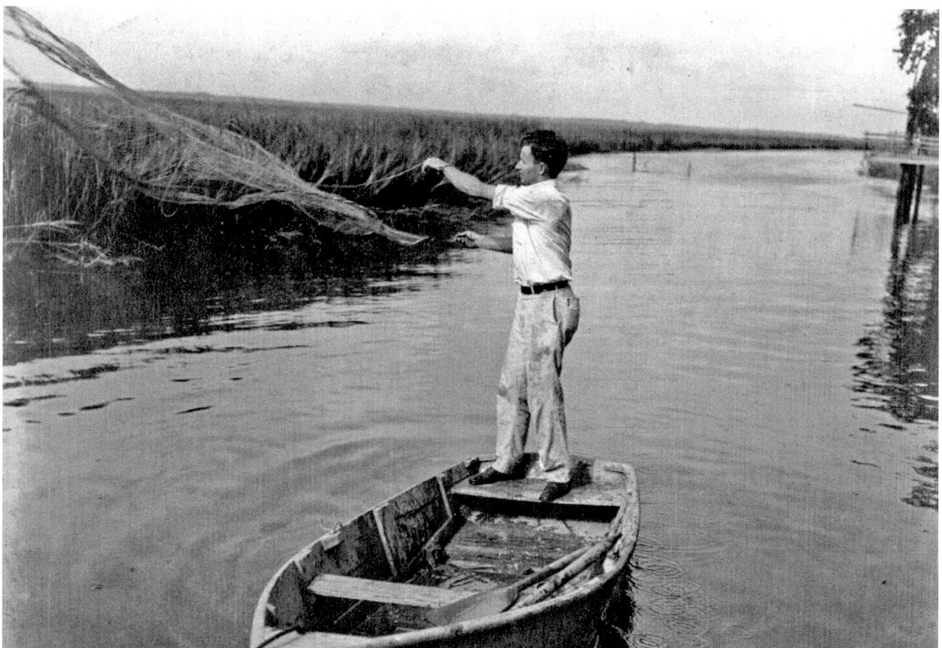

Fishing with a cast net in a local creek. *Amelia Island Museum of History Archives—courtesy of Billy Burbank III*

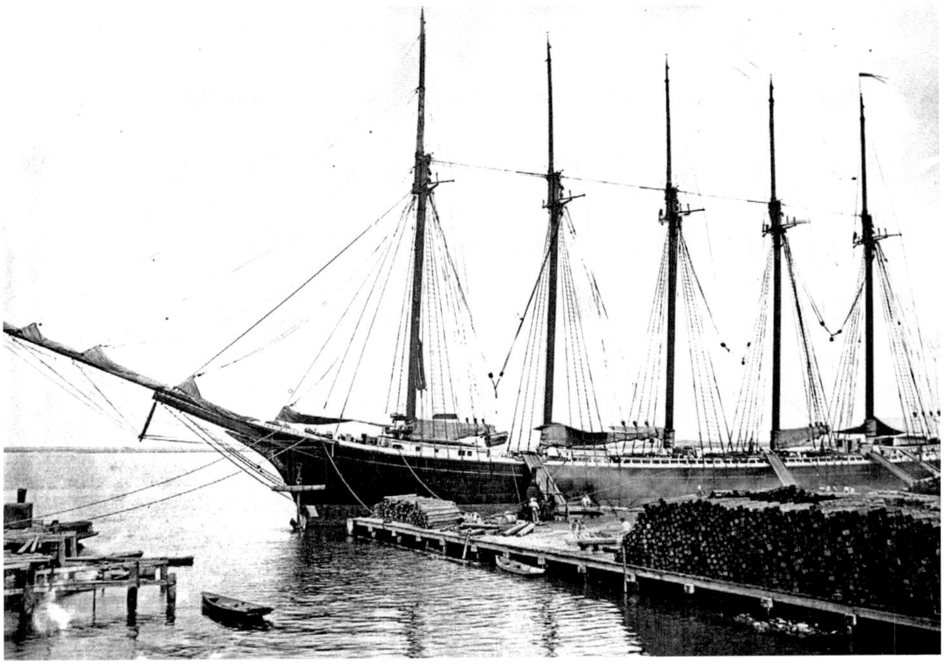

Five-mast Schooner in Fernandina's harbor. *AIMH Archives*

as part of the government's first statistical report on fisheries in 1880, he said he had been "obliged" to shut down the once lucrative pipeline "after meeting with some losses and bad returns from dealers." No one knows for sure what ended Fisher's enterprise, but its demise would initiate a lull of several decades for exporting local shrimp.[2]

People of course continued catching shrimp in Fernandina, and the number caught yearly continued to rise in the 1880s, peaking at 97,280 pounds in 1888. The local market remained capable of producing shrimp (despite a large anomalous dip in the 1890s), however without Fisher's network of distribution, there was no way to export it. While the local shrimp industry took a big step back in the last two decades of the nineteenth century, Fernandina was building up infrastructure that would help facilitate the resurgence of shrimping – then only a few decades away. Amelia Island had become one of the nation's busiest ports for the exportation of pulp and fertilizer during the late Victorian Era, and as such had extensive docks along the downtown area and well-maintained facilities. The maritime tradition was firmly in place.

The Mallory line of steam ships established weekly service between New York and Fernandina, bested by the Clyde Line who set up bi-weekly service in 1886. In 1894 the Florida Railroad was able to link up with the rest of the major lines on the eastern seaboard for the first time, essentially giving Amelia Island a direct connection with the rest of the country. The invention of the refrigerated boxcar in 1875 would be as crucial to shrimp distribution as the jet engine was to international travel. Shrimp could now be shipped further and on longer journeys than had ever been possible previously.

T. E. Fisher's shrimping operation may have been short-lived, but it set the stage for the rise of the modern shrimping industry just a few decades later. The venture demonstrated that shrimp were abundant in local waters, and set a precedent for exporting those shrimp to other parts of the country. The turn of the twentieth century would bring several people to Amelia Island who would recognize the area's full potential and find ways to realize it.

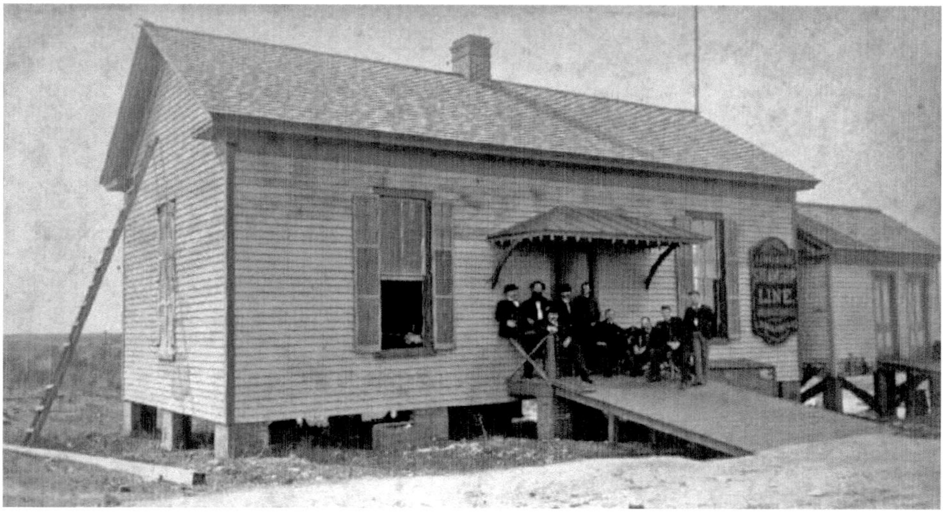

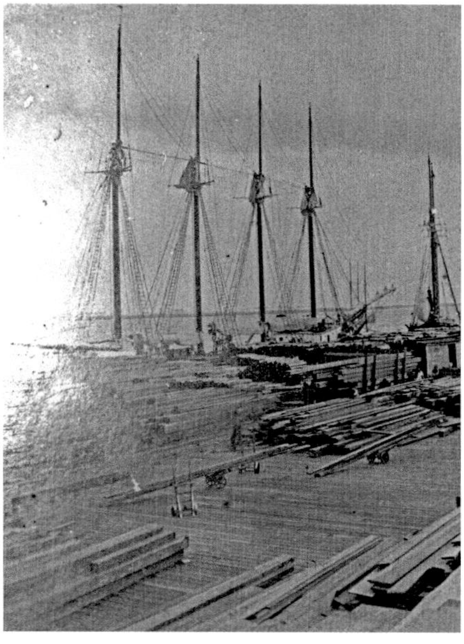

Above: Local Office of the C. H. Mallory and Co. Steamship Line. *AIMH Archives*

Left: Lumber and cross-ties at docks belonging to the Mallory docks. *AIMH Archives*

AMELIA ISLAND'S HISTORY

Amelia Island is the northern most barrier island in Florida, roughly a mile's distance from Cumberland Island, Georgia. The island is about thirteen miles long by two miles wide, and is home to the city of Fernandina Beach – the county seat of Nassau County, Florida. Amelia Island's East Coast runs out into the Atlantic Ocean. The northern tip of the island touches the St. Marys River, the southern point dips into the Nassau Sound, and the west coast borders the Nassau River.

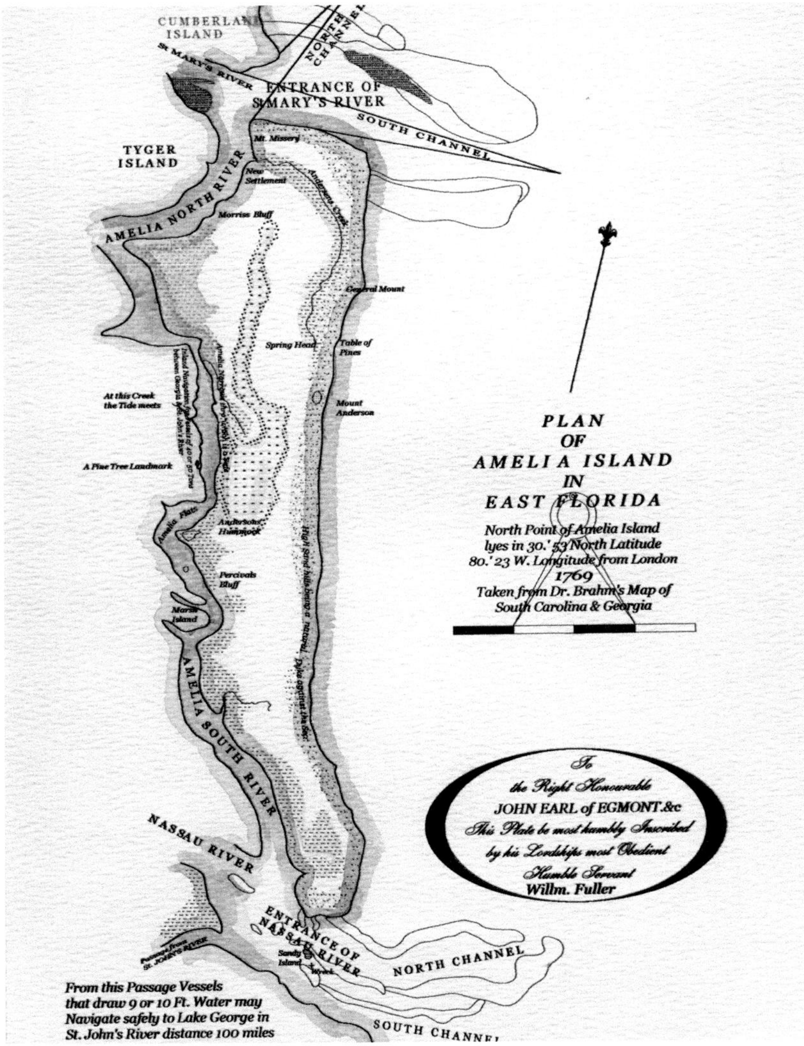

Amelia Island in East Florida charted by Captain William Fuller in 1770.

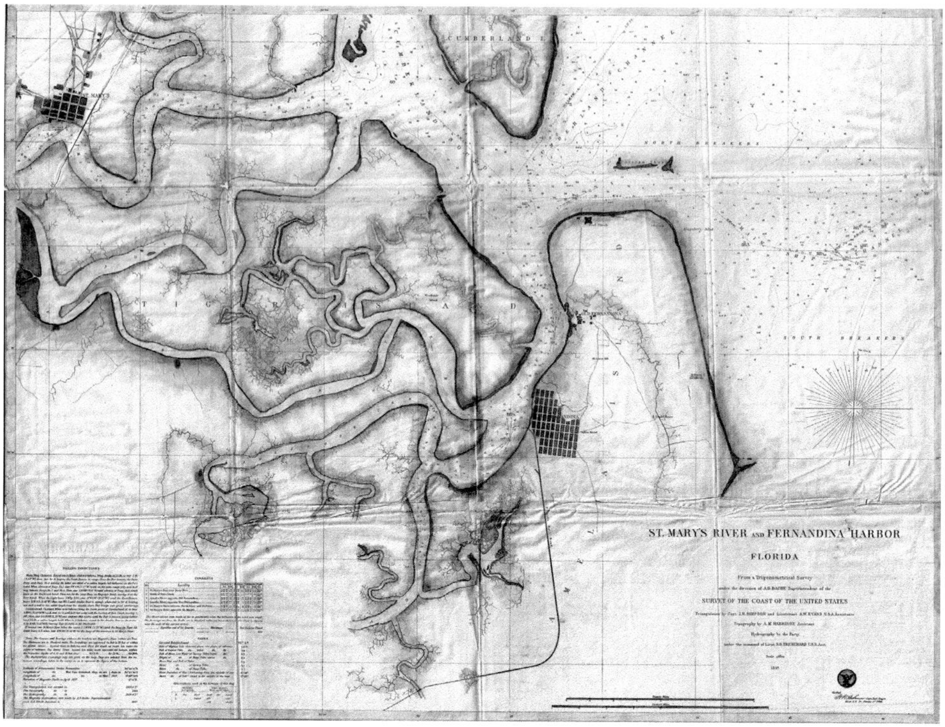

Hand colored 1857 U.S. Coast Survey sea chart or map depicting St. Mary's River and Fernandina's Harbor.

As early as AD 1000, the island was the seasonal home of Native Americans who were dubbed the Timucua by Europeans. Amelia Island is often called the "Isle of Eight Flags" in reference to eight sovereign flags that have flown over it. The island has been claimed in the name of the French (1662-1665), the Spanish (1665-1763 and 1783-1821 with interruptions), the British (1763-1783), the Patriot Flag (1812), the Green Cross of Florida Flag (1817), the Mexican Rebel Flag (1817), the United States (1821-1861), and the Confederacy (1861-1865).

The first Europeans to visit the area were led by French Explorer Jean Ribault, who came to northeast Florida in 1662 searching for a suitable location to set up a colony of French Huguenots. He called Amelia Island *Île de Mai* (the Isle of May in English). In 1664 René Goulaine de Laudonnière established Fort Caroline on the banks of the St. Johns River. The Spanish considered Florida part of their territory in the New World, and would not countenance having French Protestants settled there. Sent by Phillip II of Spain, Pedro Menéndez de Avilés founded St. Augustine in 1865 and attacked Fort Caroline, eradicating the French presence in Florida. Amelia Island was part of the Spanish Mission system and two missions were established on the island at different times during Spanish rule, Santa Maria de Sena, and later Santa Catalina De Guale.

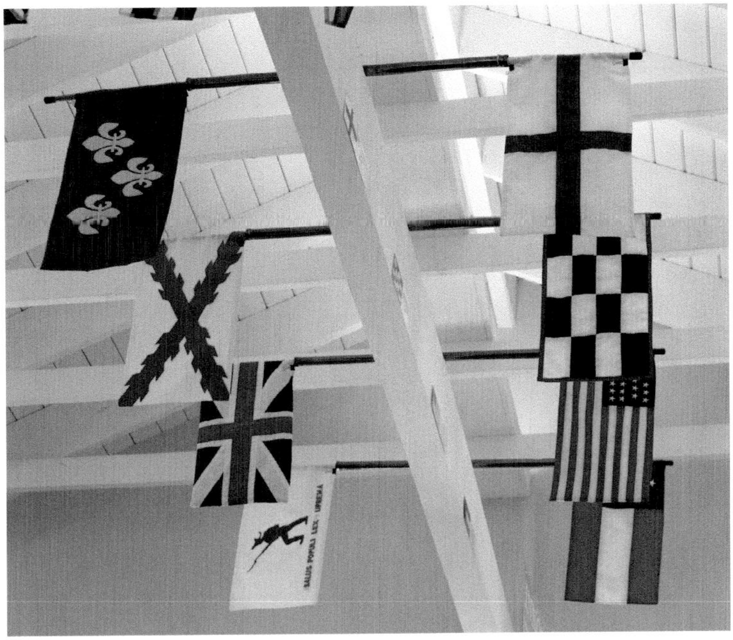

The 8 Flags of Amelia Island displayed at the Fernandina Beach Shrimping Museum and Welcome Center.

In the late 1600s the English – led by South Carolina Governor James Moore – raided Spanish missions in Florida including Santa Catalina, which was burned to the ground. James Ogelthorpe, founder of the Georgia colony, negotiated with Spanish officials to make the island a British possession during the War of Jenkin's Ear (a conflict between Great Britain and Spain which began in 1739, and was later subsumed into the War of Austrian Succession), but the King of Spain refused. It was Ogelthorpe who gave the island the name Amelia, in honor of King George II's second daughter.

The 1763 Treaty of Paris brought an end to the Seven Years War and as part of the treaty Britain acquired Florida from Spain in exchange for Cuba. Under British control, the Florida colony was separated into two separate territories, East and West Florida – each with its own capital and governor. The British Crown granted John Perceval, the Second Earl of Egmont, 10,000 acres on Amelia Island. Perceval's holdings became a large indigo plantation run by Superintendent Stephen Egan. As the Revolutionary War wound down in the late 1770s, Amelia Island became an evacuation point for loyalists fleeing the southern colonies.

The Second Treaty of Paris, in 1783, returned Florida (now divided into two territories, East and West Florida) to Spain. Fernandina was first platted by the Spanish as part of East Florida in 1811. Named after Spanish King Ferdinand VII, the town was platted by surveyor George J. F. Clarke. The original center of town (known to locals as "Old Town") was about a mile north of what is now considered Fernandina's historic district. The Spanish erected a small wooden fort overlooking the Amelia River in 1816.

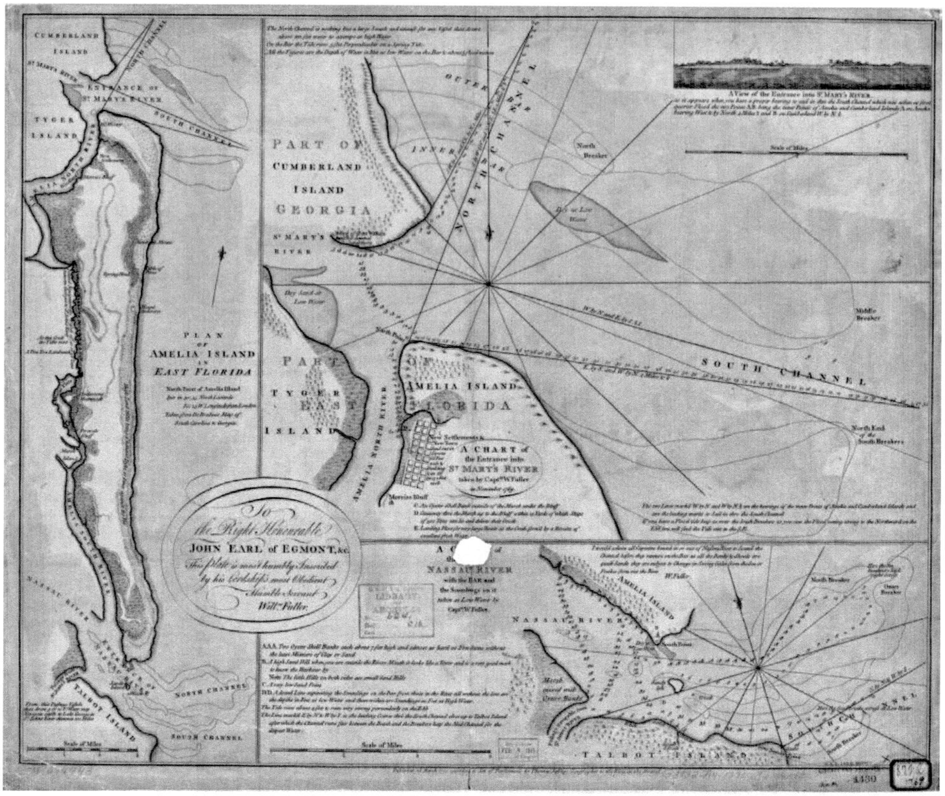

1777 Chart of the entrance to the St. Mary's River including Amelia Island. *Library of Congress, Geography and Map Division*

The Embargo Act of 1807 – which prohibited the United States from trading with Great Britain and France – for a while made Fernandina one of the busiest ports in the world, and a haven for smugglers. Amelia Island was a Spanish possession conveniently close to Georgia, which made it an ideal location for smuggling contraband goods into the United States. Amelia Island's proximity to the United States was tremendously appealing, and several different parties would try to take possession of it in the years to come.

On March 13, 1812, a group of insurgents led by John McIntosh, a planter with land in both Georgia and East Florida, invaded Fernandina backed up by US gunboats. They were masterminded by American Revolutionary war hero George Matthews, who probably had the implicit blessing of President James Madison. Calling themselves the "Patriots" and hoping to incite rebellion against the Spanish in East Florida, they raised their own flag over Amelia Island, which was replaced by the United States flag the following day. The "Patriots" marched on Saint Augustine, the most significant settlement in Florida, and unsuccessfully laid siege to the Castillo de San Marcos. The incursion failed and the

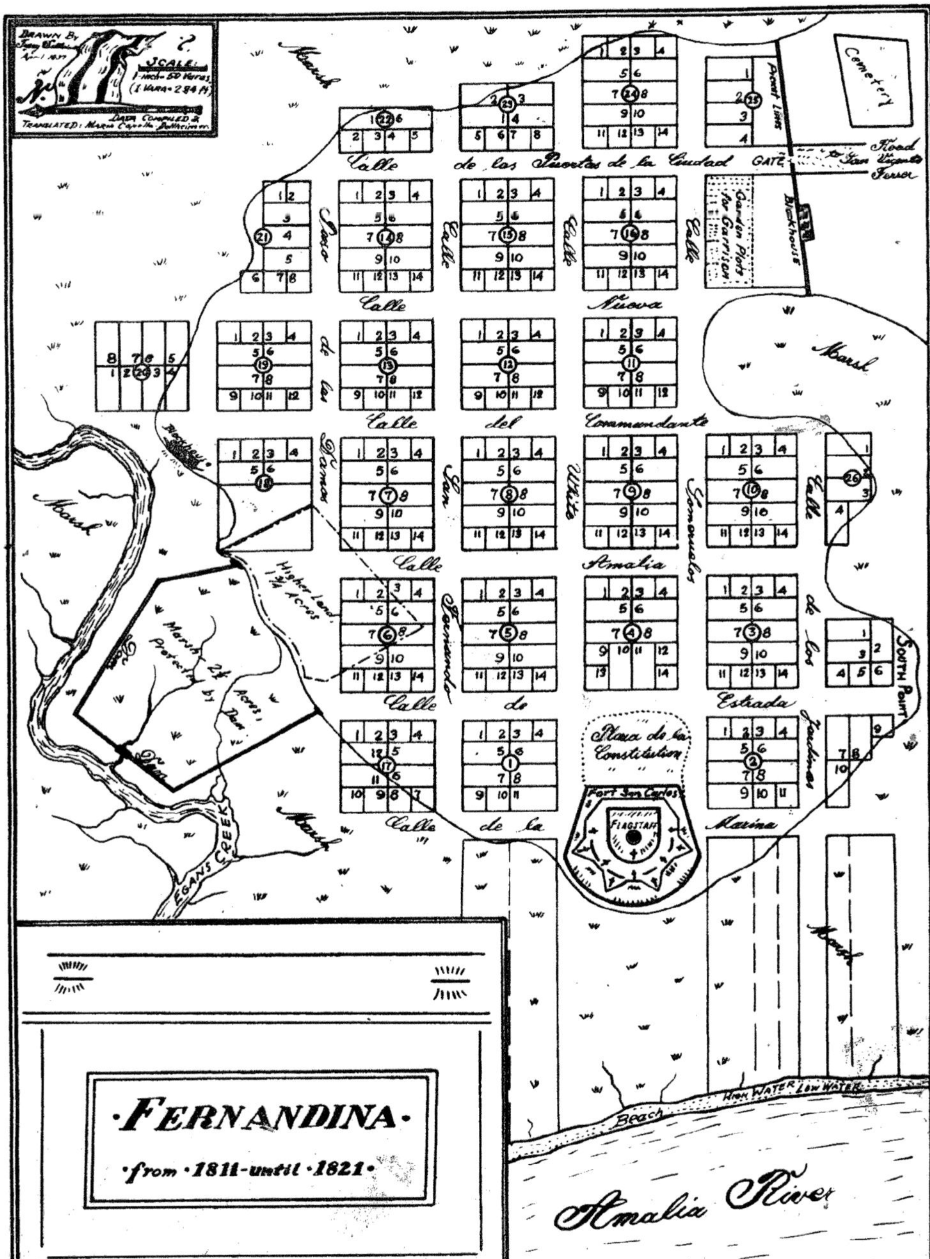

Fernandina as it was platted by surveyor George Clarke in 1811. British by birth, Clarke became a loyal Spanish subject and one of the most important men in East Florida during the Second Spanish Period. *AIMH Archives*

Spanish regained control of Amelia Island. The United States, then on the brink of war with Great Britain, could not afford to make an enemy of Spain. President Madison disavowed any knowledge of the plot.

Spain's control of Amelia Island remained tenuous, and the smuggling of goods and slaves into nearby Georgia was still prevalent. In the summer of 1817, Scottish mercenary and adventurer Gregor MacGregor, who claimed he represented the interests of various South American countries, raised a small, privately funded army with which he planned to invade Florida – with Amelia Island as his first target. With around fifty to a hundred men MacGregor marched on Fort San Carlos and took the fort without a shot, raising what historians call the "Green Cross Flag." MacGregor set up an ad hoc government and waited for the reinforcements and supplies he had supposedly been promised to complete his mission. By early September, MacGregor was out of money and his men were growing restless. He abandoned Amelia Island, leaving it in the hands of American expatriates Ruggles Hubbard and Jared Irwin.

Hubbard and Irwin (with ninety-four men) successfully repelled a Spanish attempt to take back the island on September 13, 1817. Four days later, French privateer Luis Aury, fresh from harassing Spanish ships in the Gulf of Mexico, sailed into Fernandina and dubiously annexed Amelia Island on behalf of the Republic of Mexico. Growing frustrated with the chaotic situation on Amelia Island, President James Monroe decided to intervene. Aury surrendered control of the island to US Forces in December 1817, with President Madison vowing to "hold it in trust" for Spain.

As part of the Adams-Onis Treaty, Spain ceded both East and West Florida to the United States in 1819, and Florida became the twenty-seventh state to join the United States of America. The United States flag became the Florida Statesman and businessman David Yulee brought the railroad to Fernandina in the 1850s, making the town the east coast terminus of his Florida Railroad. Yulee convinced most residents to move the center of town approximately one mile south, to its present location.

Florida seceded from the Union on January 10, 1861. Two days prior, local Confederate militia had taken over the abandoned (and unfinished) Fort Clinch, which the US Army had begun constructing in 1847 during the Second Seminole War. Confederate leadership decided Amelia Island was strategically not worth defending and ordered that it should be evacuated in the case of a Union offensive. On March 3, 1862 Commodore Samuel DuPont sailed into Fernandina with twenty-eight gunboats and reestablished federal control over the island. Fernandina became a haven for escaped slaves and freedmen, many of whom were recruited into the 1st South Carolina Volunteers – the first black regiment in the United States Army.

After the war, Fernandina entered what local historians now refer to as "the Golden Age of Amelia Island." Many of the Union soldiers who had been stationed on Amelia Island during the war returned to settle with their families. With more people came an influx of commerce and capital. Fernandina's geographic location and harbor once again showed its appeal, as the city became one of the most important ports in the nation for the exportation of phosphate and lumber.

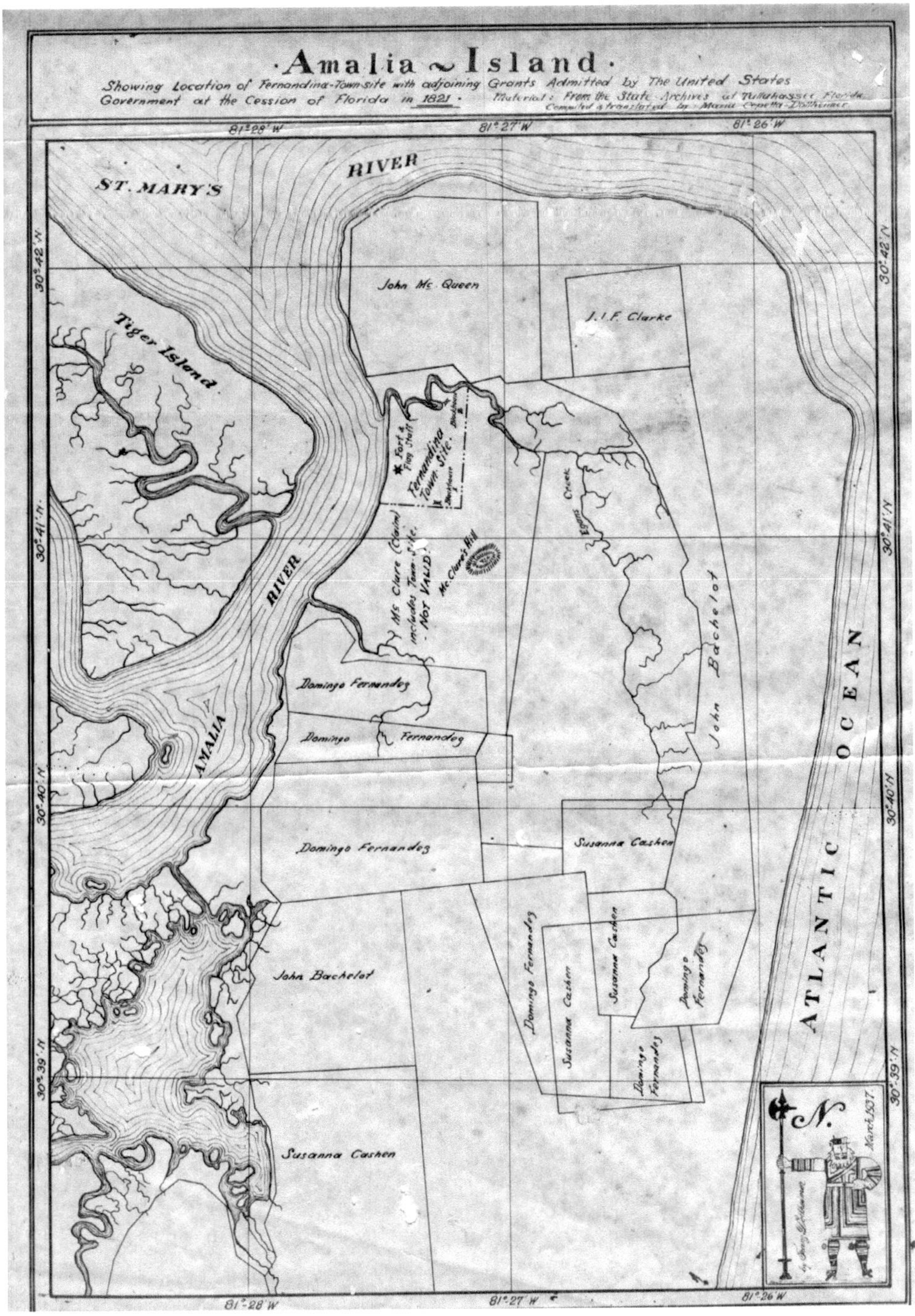

1821 Map of Fernandina. *AIMH Archives*

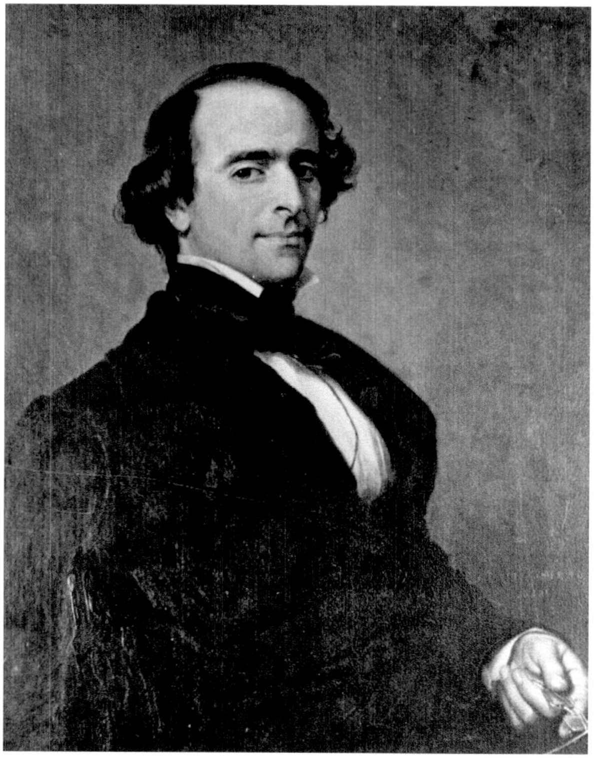

Left: Portrait of David Levy Yulee (1810-1886) was an attorney, planter, entrepreneur, and statesman, who was instrumental to Florida's Statehood, and its development via the Florida Railroad. *AIMH Archives*

Below: 1857 map of Fernandina. David Yulee envisioned Fernandina as the "Manhattan of the South," and had the town plotted accordingly (complete with its own Central Park). *AIMH Archives, photo by Christopher Collins*

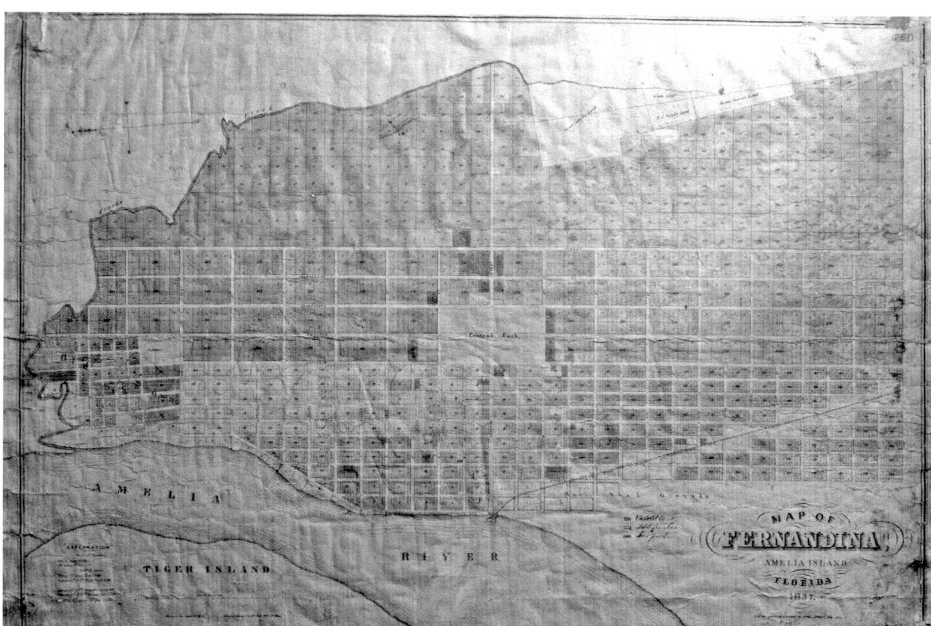

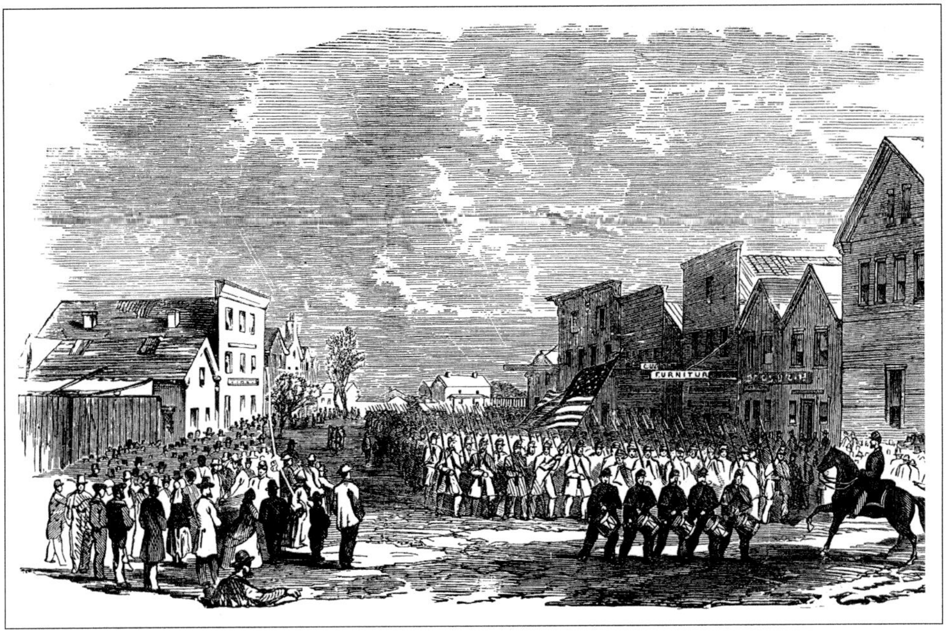

Federal troops march down Second Street in Fernandina, Florida. Taken from Frank Leslie's *Famous Leaders and Battle Scenes of the Civil War*, 1896. *AIMH Archives*

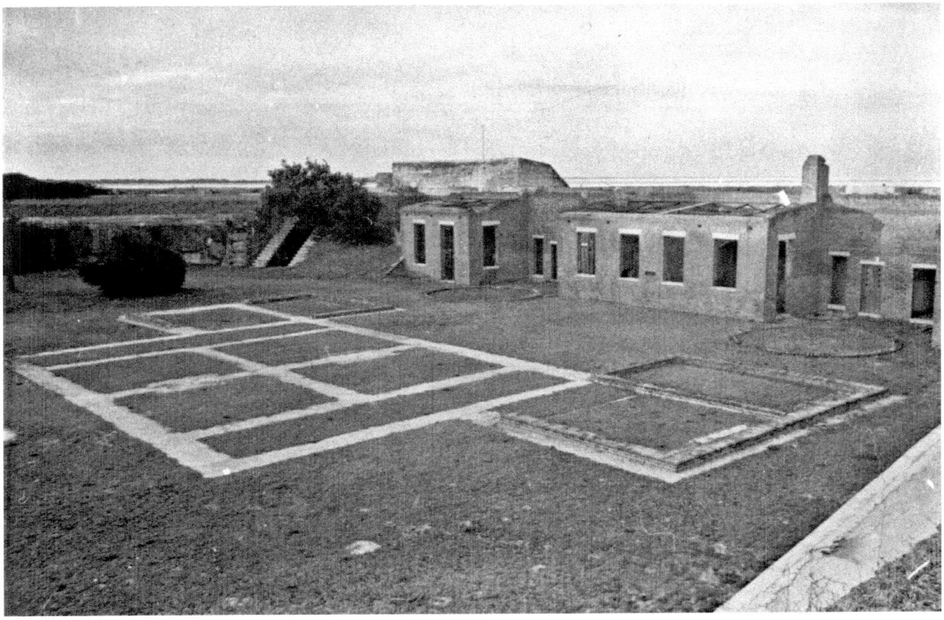

Interior view of Fort Clinch before the Civilian Conservation Corps' restoration in the 1930s. *AIMH Archives*

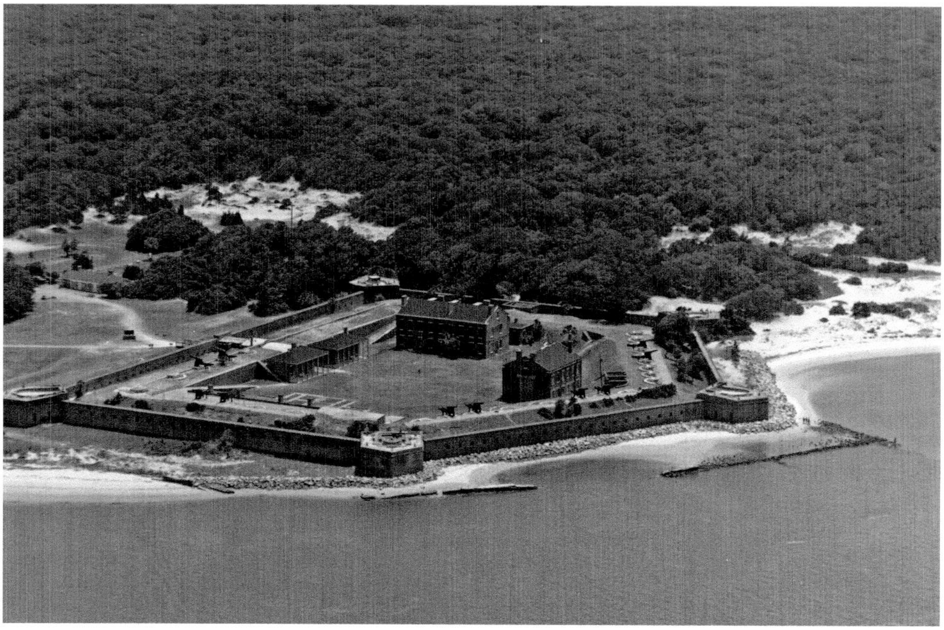

Aerial view of Fort Clinch. *AIMH Archives*

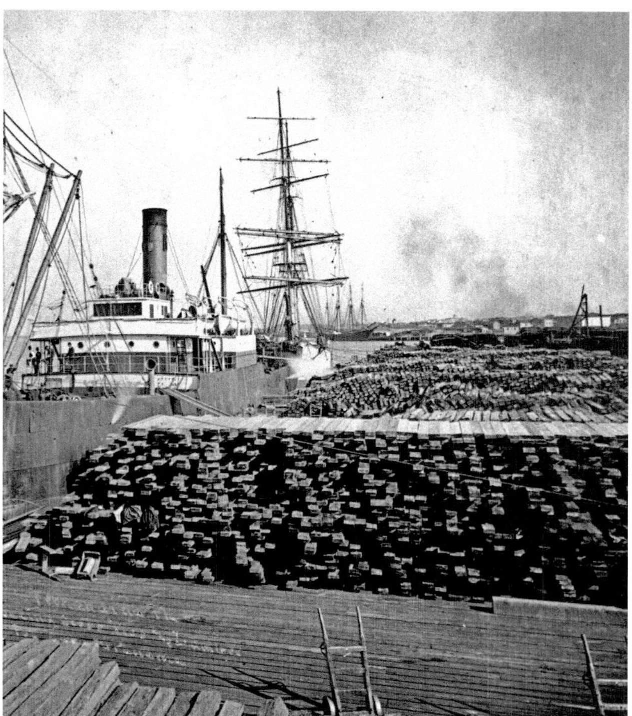

Loading lumber at Fernandina. *AIMH Archives*

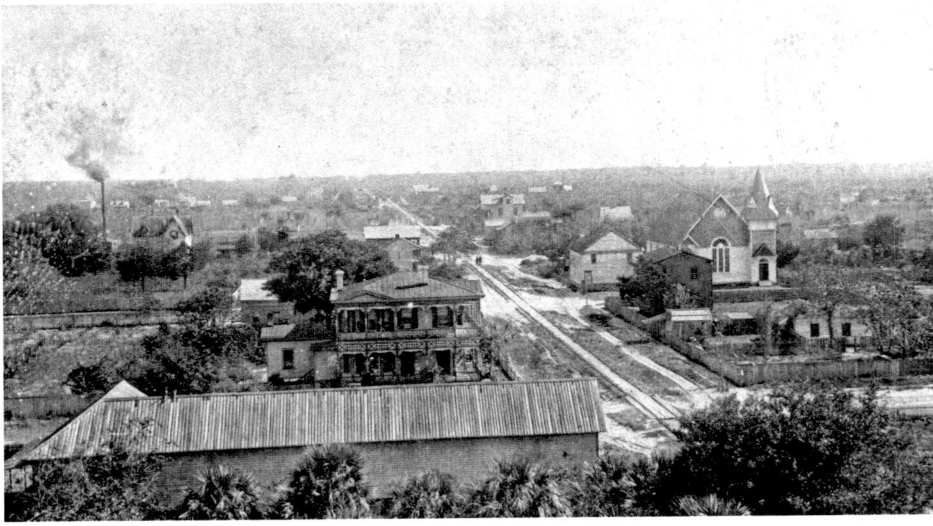

Downtown Fernandina in the Victorian Era, taken from the Egmont Hotel. *AIMH Archives*

For a while, Amelia Island was a fashionable tourist destination. Wealthy families from the north flocked to northeast Florida to vacation. Several hotels sprang up on the island, like the Florida House Inn (opened in 1857), the Strathmore Hotel (which followed in 1881), and the Egmont (built in 1878) – by far the most extravagant. By the early twentieth century, tourists began moving south, and Fernandina's reputation as a resort destination faded. The Strathmore was destroyed by the Hurricane of 1898, and The Egmont Hotel, with its sixty-five rooms, billiards rooms, bowling alley, and croquet grounds, proved to be too grand for its own good. There were simply not enough guests in town to fill so many rooms. After hemorrhaging money for several years, it was demolished in 1906.

Amelia Island, with all its maritime potential, was in danger of withering economically. In the 1930s two pulp and paper mills would come to town bringing jobs and people with them. But before that measure of stability arrived, there were several decades where the island's future looked bleak. A new industry was needed to somehow breathe life into the fading city. A confluence of Amelia Island's natural potential along with the introduction of the right people (hard workers with an entrepreneurial spirit) would reinvigorate the local economy and allow the island to reinvent itself once again.

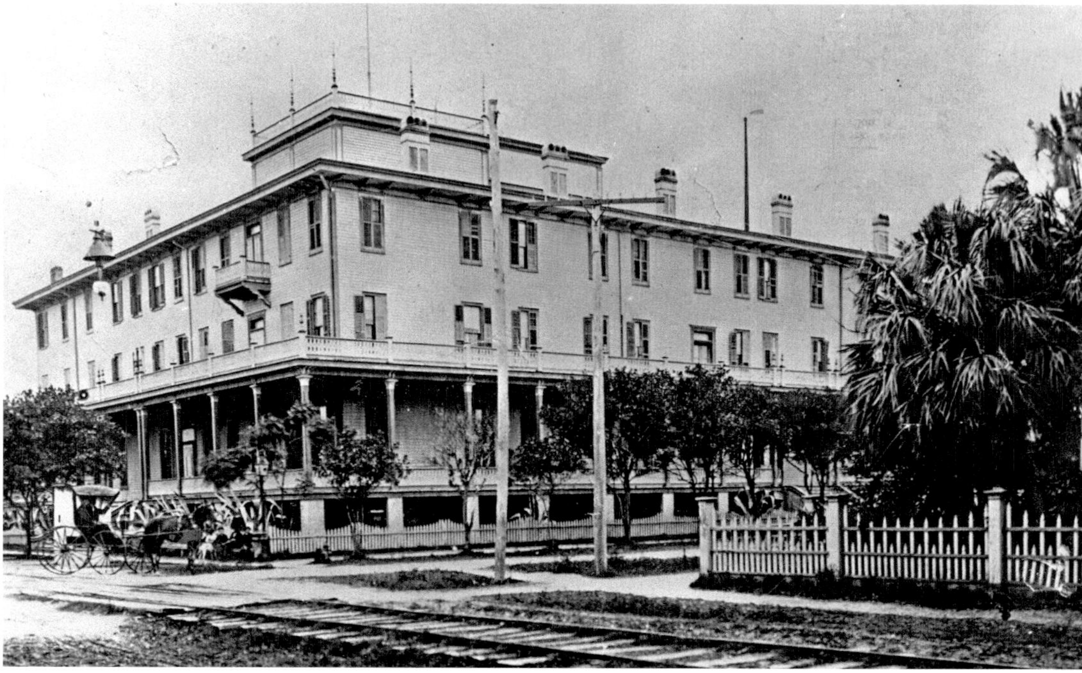

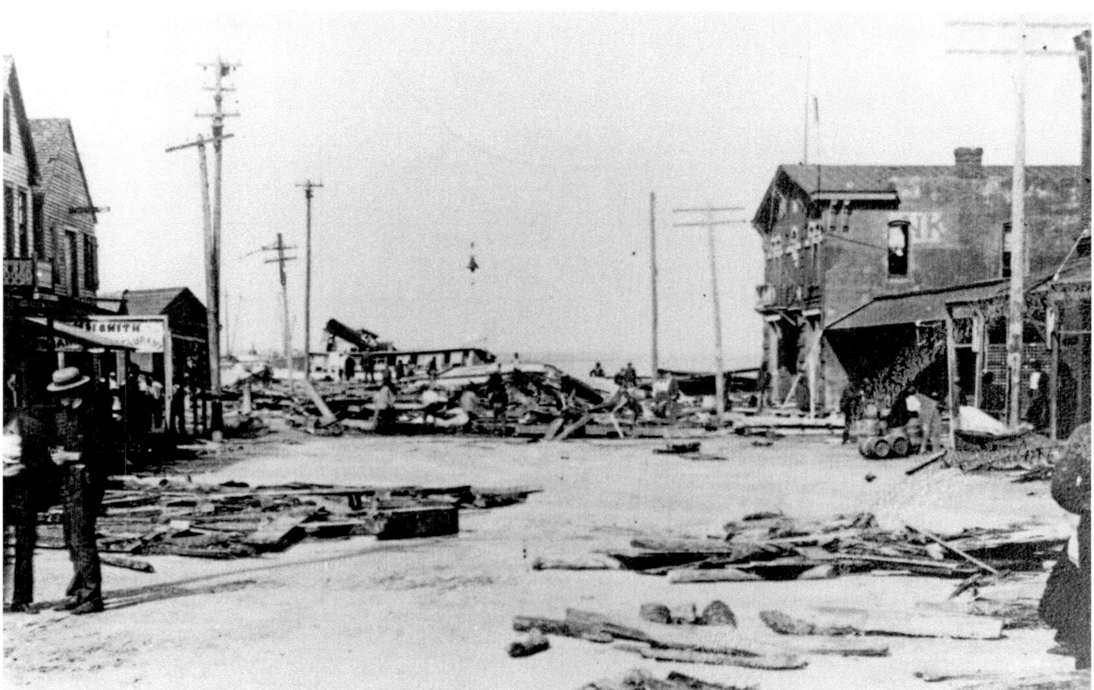

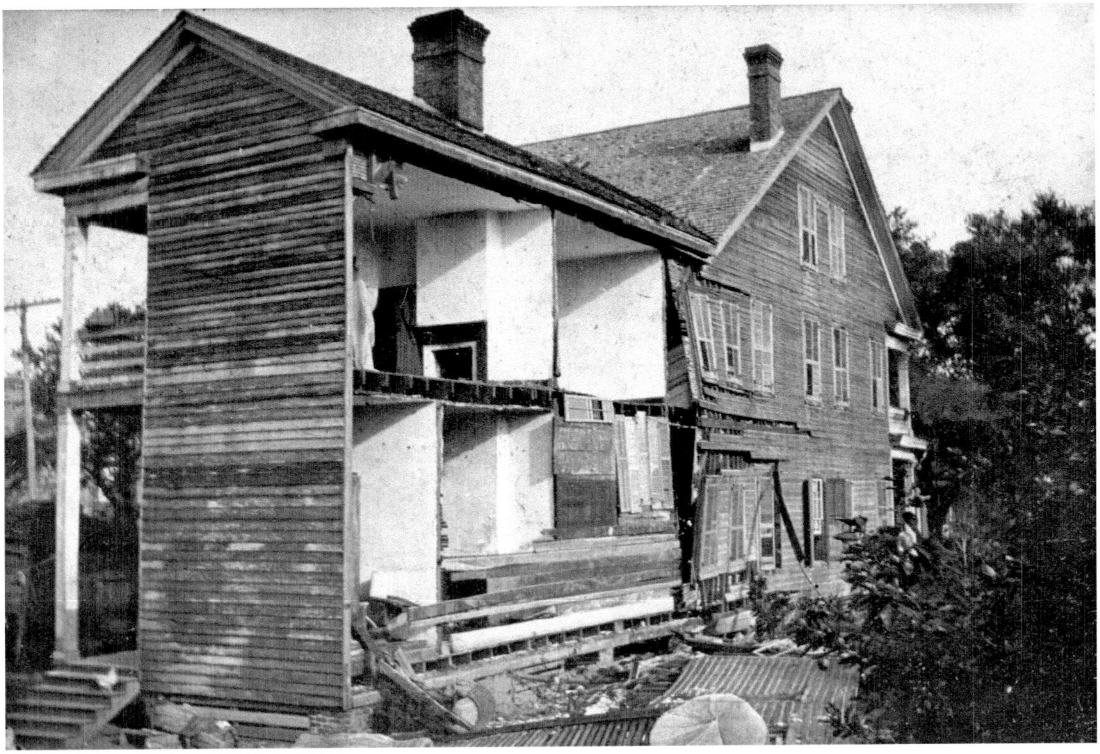

Home destroyed by the Hurricane of 1898. *AIMH Archives*

OPPOSITE PAGE:

Above: The opulent Egmont Hotel. *AIMH Archives*

Below: Damage from the Hurricane of 1898. This photo was taken looking down 2nd Street from the Huot building. *AIMH Archives*

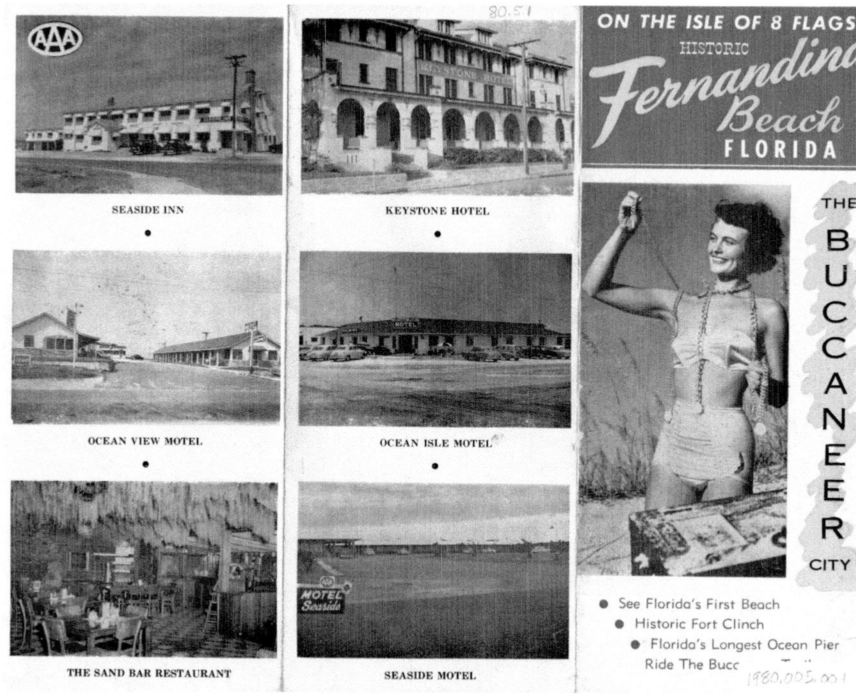

Brochure for Historic Fernandina Beach (front above, back below). *AIMH Archives*

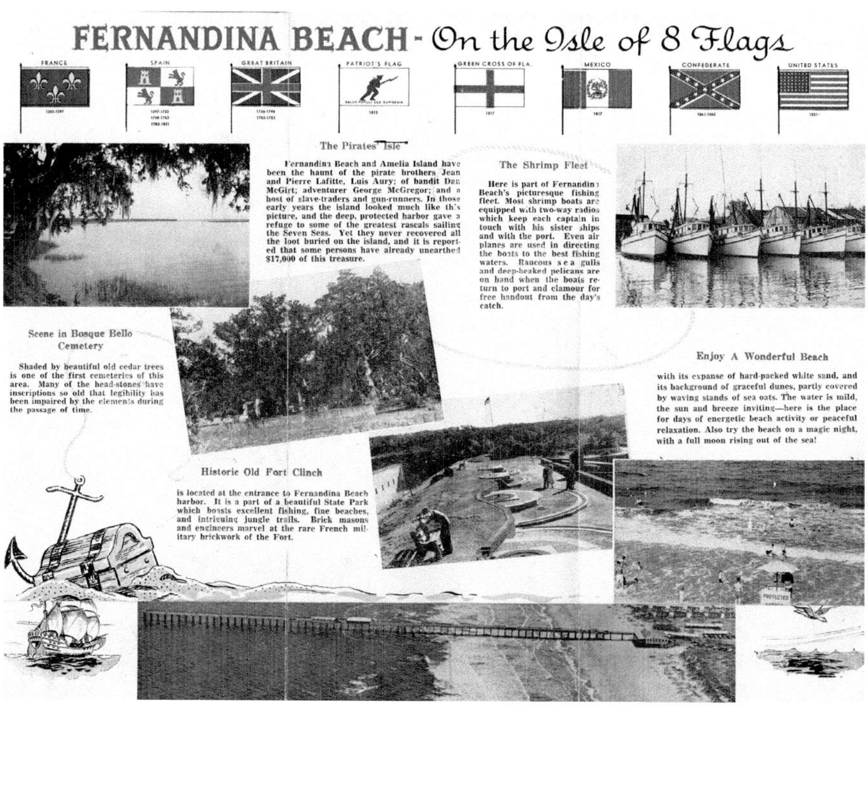

1

The Pioneers

In 1898 an Italian immigrant named Sollecito Salvador came to Fernandina. Salvador would leave his mark on shrimping, not only by pioneering one of the biggest technological innovations in the history of the industry, but also by convincing several of his friends from Sicily to join him on Amelia Island. Through a series of marriages, these relocated Sicilians would become a large, extended family. The resulting familial ties led to cooperation in business and sharing of assets which helped get the shrimping industry in Fernandina off the ground in its earliest days.

Sollecito Salvador was born in 1869 in Syracuse (Siracusa in Italian), a historic province on the island of Sicily. At that point in time the island of Syracuse was becoming crowded and opportunity was scarce. Searching for a place to make a life for himself, Sollecito joined the merchant marine as a young man and made his way to America. His path to Fernandina has two conflicting origin stories. The local lore is that Salvador was on an Italian freighter that was forced to dock in Fernandina by the savage Hurricane of 1898.

However, his nephew, Brother John Versaggi SDB, records that Salvador first arrived in the US by way of New Orleans in 1895, settling for a time on the west coast of Florida, in Cedar Key. While visiting Amelia Island by train, he saw potential in the local fisheries and decided to relocate to Fernandina. Family tradition holds that he was actually born Salvador Sollecito, but his name was somehow inverted by mistake during the immigration process. Not wanting to jeopardize his entrance into the US, he simply adopted the change.

However he got here, both sources agree that Sollecito saw enough potential in local fisheries to settle down and make Amelia Island his home. Salvador – who took the name "Mike" because locals had difficulty pronouncing Sollecito – made his living as a fisherman and an interpreter for Mediterranean ships that came to Fernandina. Working part time on other shrimper's boats he was exposed to the early technological innovations swirling around Fernandina.

Salvador saved up enough money to purchase his own boat, along with equipment for drying and preserving shrimp. A chattel mortgage registered June 23, 1899 records his acquisition of the tools he would need to get his start in the shrimping business, including: "one large ice box, one small ice box, one ten-gallon iron boiler, one shrimp cast net, twenty

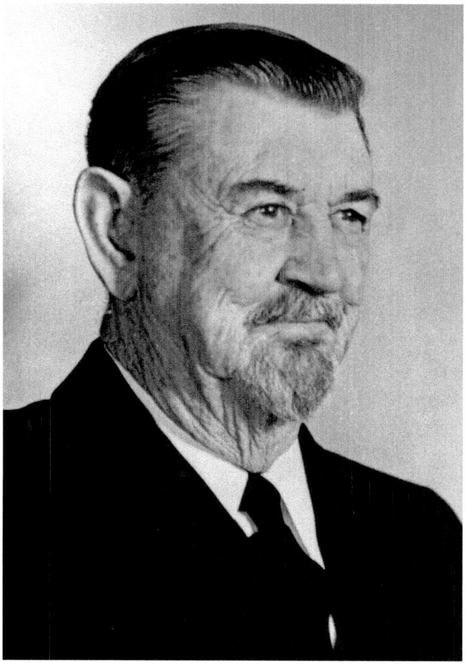

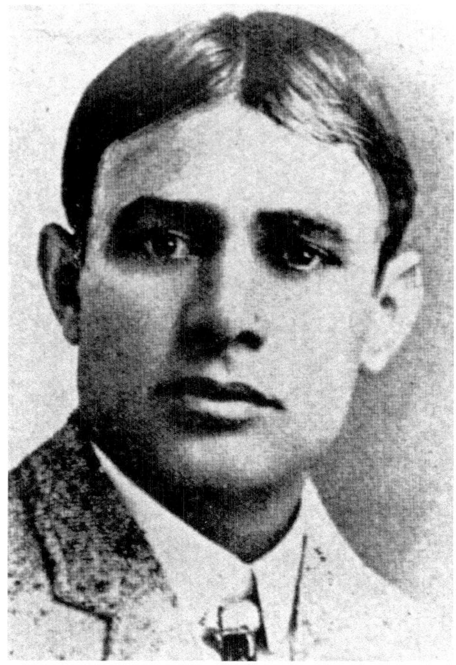

Left: Master Pilot William Jones Davis. *AIMH Archives*

Right: Sollecito "Mike Salvador" as a young man. *AIMH Archives*

kegs, and 150 pounds of preserving powder." Salvador made one of the most significant contributions to modern day shrimping when he introduced motorized boats to the industry sometime around 1902. Powered boats could go further away from shore into deeper water, and come back with a substantially larger catch than shrimpers using traditional methods.

Local harbor pilot Captain William Jones Davis is credited with being the first to attempt fishing off shore, another shift that would revolutionize the industry in the coming years. Davis equipped his boat the "Automatic" with trawls he intended to use for catching flounder. One day, as he was dragging for flounder in deep water off the coast of Amelia Island, the trawl came up with a remarkably large shrimp catch. Davis' find inspired shrimpers to start trawling off shore, where they were able to catch substantially more shrimp. Boats had to grow larger to match production. Shrimpers needed more horsepower to pull bigger nets, and more room in the hold to store the extra shrimp they were catching.

The shift in the industry was rapid. Shrimpers were quick to adapt to new methods and equipment. Haul seines were subbed out for Otter Trawls, and rowboats were replaced with fleets of powerboats. The biggest difference between the late Victorian era local shrimping industry of T. E. Fisher's day, and this fresh rebirth was not only in the equipment, but the sheer scale. By 1915 Salvador had a flotilla of motorized shrimp boats including the

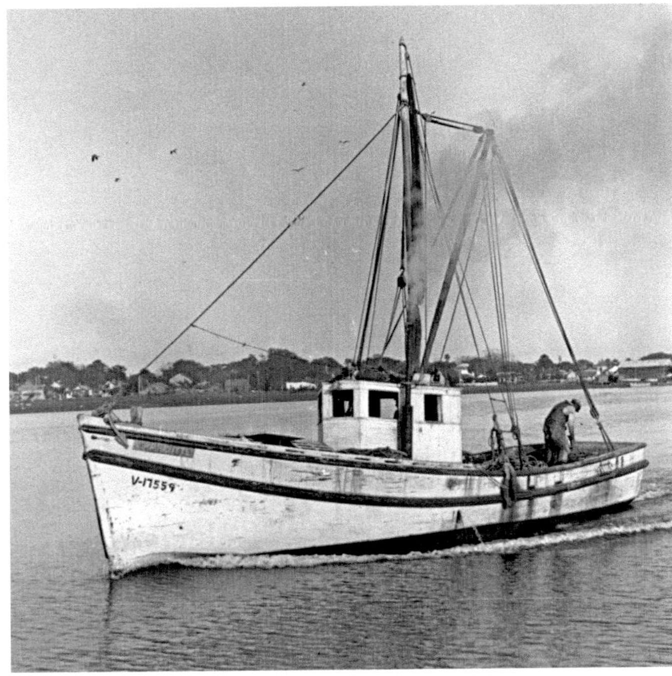

The *Charlotte* trawling near Fernandina. *AIMH Archives*

"Domenica," the "Amelie," a forty-six foot launch built by local builder Oscar Danburg, a 14 ton, 53 foot vessel, and the 26 foot 9 inch "Lucy S," with many others to follow.

In 1906 Salvador started the Salvador Fish company (later known as S. Salvador and Sons), prominently located on Centre Street across from the Palace Saloon (claimed to be the longest continually operating Saloon in Florida). Sending letters back to Italy extolling Amelia Island's weather and ample fisheries, he was able to convince his friends Salvatore and Antonio Versaggi to join him in Fernandina. Leaving from Augusta, Italy, in 1905 the brothers were joined by Antonio's wife Concetta, their sister Carmela – along with her husband Antonio Poli – and their daughter Josephina. Salvatore's wife Vincenzza Litrico followed after giving birth to a son named John that same year. In 1907 their younger sisters Virginia (with her husband Joseph Tringali) and Domenica made the trip as well.

The youngest sister, Domenica, had been trading letters with Mike Salvador for over a year, and came with the intention of becoming his wife. Upon realizing Mike was almost twice her age, Domenica got cold feet. But Mike was persistent, and with the encouragement of her brothers and sisters he convinced Domenica to marry him after two years of courtship. Family tradition holds he would bring her a rose every day, and could be heard rowing down the Amelia River singing an Italian love song that translated to "You are My Rose." The couple married in 1909 and had four children: Lucy, Nellie, Felix, and John. Mike moved his family to a two story home in downtown Fernandina, on South Fourth Street. Tony Versaggi moved in next door.

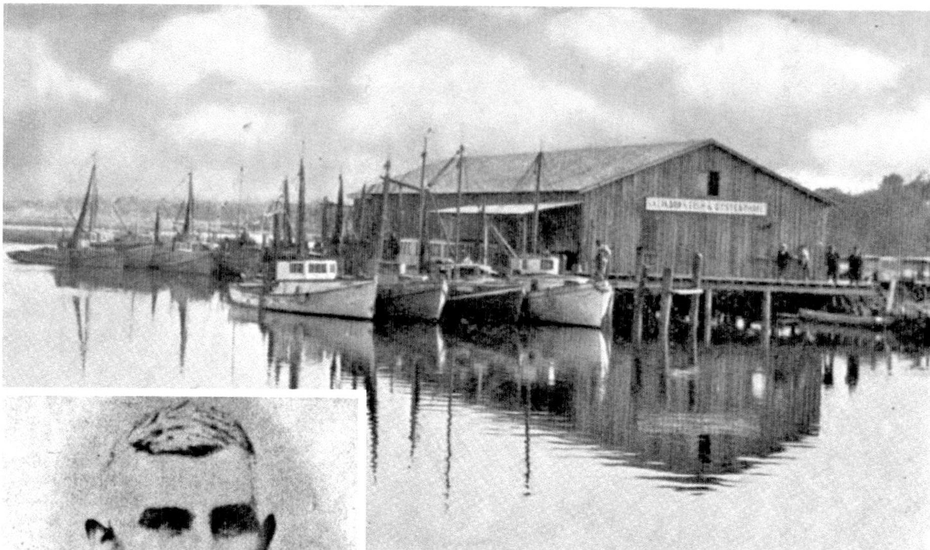

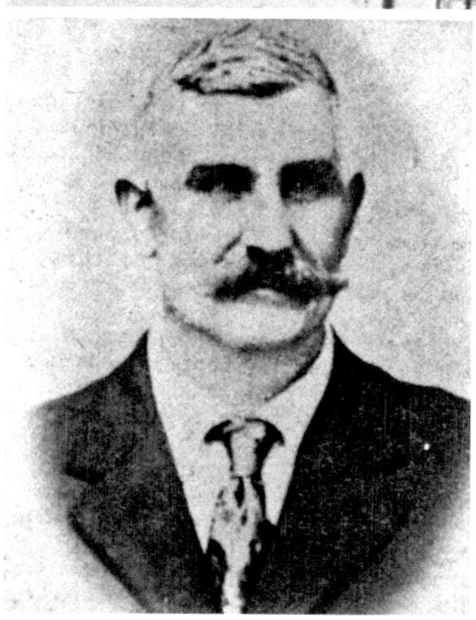

Above: Postcard of Salvador's fish house in the early 1900s. *AIMH Archives*

Left: Salvatore Versaggi. *AIMH Archives*

Below: Downtown Fernandina in the early 1900s, Salvador's Fish Market is on the left. *AIMH Archives*

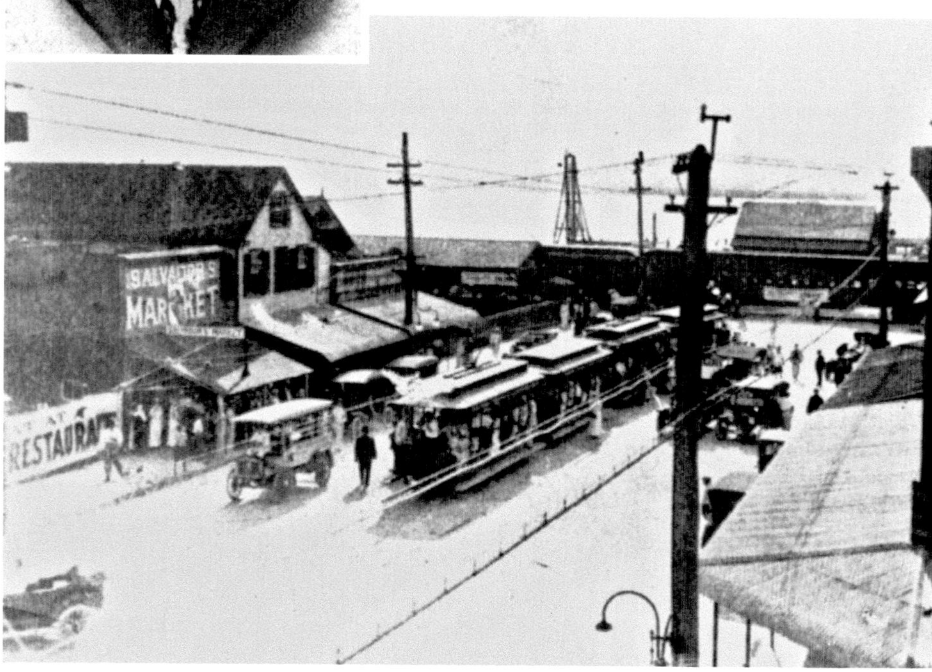

The market for shrimp at this time was soft, usually selling locally at a few cents per pound. The Versaggi brothers returned to New York City and worked as longshoremen, building important contacts with the Fulton Fish Market. Traveling back and forth between New York and Fernandina, Salvatore Versaggi was able to make a deal with Fulton for the distribution of Fernandina shrimp. When he first began shipping shrimp north, the proceeds barely covered his expenses. Shrimp being a perishable commodity, the logistics of shipping it and still turning a profit were complicated. The Fernandina shrimpers overcame the difficulty by supplying shrimp in bulk, and by engineering a new market.

Up until that time, shrimp had widely been considered a delicacy of the rich. Versaggi and Salvador convinced Italian saloons to offer shrimp to their customers to help build a taste for the product. Demand for shrimp by the general public rose steadily throughout the 1910s and 1920s. Shrimping season began near Fernandina in the summer, and moved up the east coast to North Carolina. Local processing plants would take the shrimp from the boats, clean, pasteurize, and put the shrimp on ice to be shipped by train to Savannah, Philadelphia, and New York City. Around the same time, another Italian named Joseph DeGrande pioneered shipping shrimp by truck. The network for shrimp distribution pioneered by the Fernandina dealers set the template that is still used today.

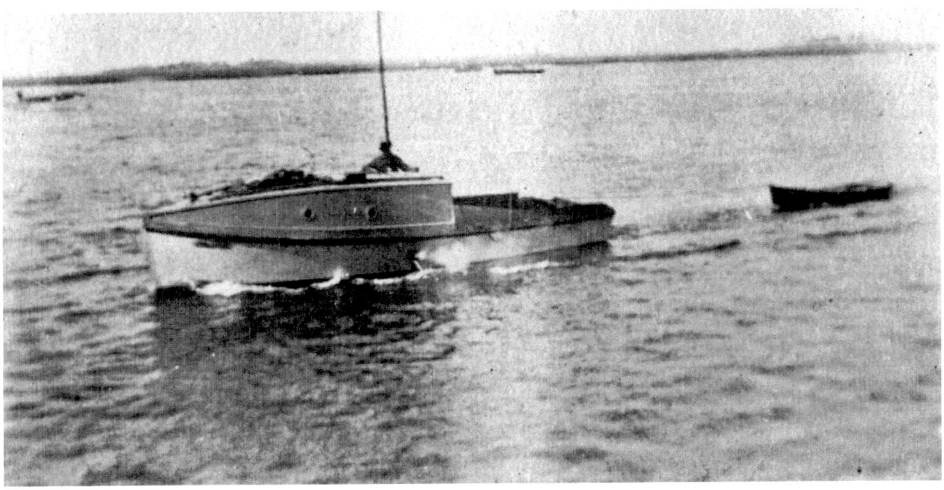

The *Renee Goddard*, an early shrimp boat owned by local entrepreneur L. G. Hirth. *AIMH Archives*

2

The Boom and Departure

Mike Salvador, the Versaggi brothers, Antonio Poli, and another Sicilian named Joseph Gianino worked together closely, pooling their resources. They shared fish houses, equipment, workers, even the captains that ran their boats. One of these who was especially successful was a Portuguese immigrant by the name of John Santos Carinhas. Carinhas was an experienced seaman who came to Fernandina in 1915 and became one of Salvador and Versaggi's best captains.

Early motorized shrimp boats were roughly twenty to thirty feet long, and powered by crude gasoline engines which had to be cranked by hand. Fuel was crude and highly combustible, which could make shrimping an extremely dangerous business. A boat cost around $675 and was usually equipped with a two cylinder engine putting out fifteen horsepower. Nets could be up to sixty feet across, and on average would sell for about thirty-three dollars.

Fernandina was responsible for another industry innovation via the application and adaptation of the otter trawl for shrimping. Patented in England in 1894, the otter trawl is a bag-like net with weighted doors that hold the net open as it drags along the ocean floor. After the turn of the century, different trawl net designs began to develop independently in many US coastal locations. Captain Billy Corkum, who followed a run of Blue Fish to Fernandina in 1912, is credited with introducing the otter trawl to the local shrimping industry. Originally from Massachusetts, Corkum developed a design based on nets used in his native New England to catch halibut and other bottom fish. His design was widely copied, and by 1918 it had become standard equipment for shrimpers.

Around the same time Salvatore Tringali, a net maker for the Salvador and Versaggi fleets, began to experiment with a similar design by adapting sardine nets from his native Sicily. Tringali's net was smaller than the standard trawl (only 3 feet long and 12 inches high), and was especially well suited for catching shrimp. Locals Dave Cook and Emmet Freeman refined the trawl by adding corners and wings. Local innovations helped lead shrimpers world-wide to abandon the traditional haul seine nets, in favor of trawl-based designs.

By 1912 shrimping was the most important industry in Nassau County. Demand rose drastically and local shrimp production increased a staggering two thousand fold between 1897 and 1912. Over a hundred boats employing about 250 men operated out of Fernandina

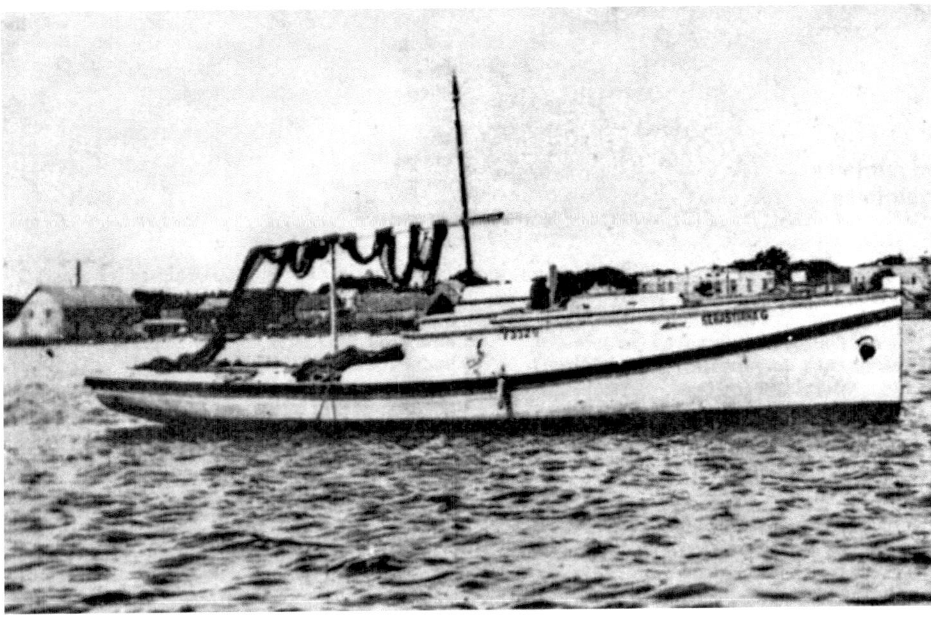

Joseph Gianino's boat the *Sebastiana* in 1915. *AIMH Archives*

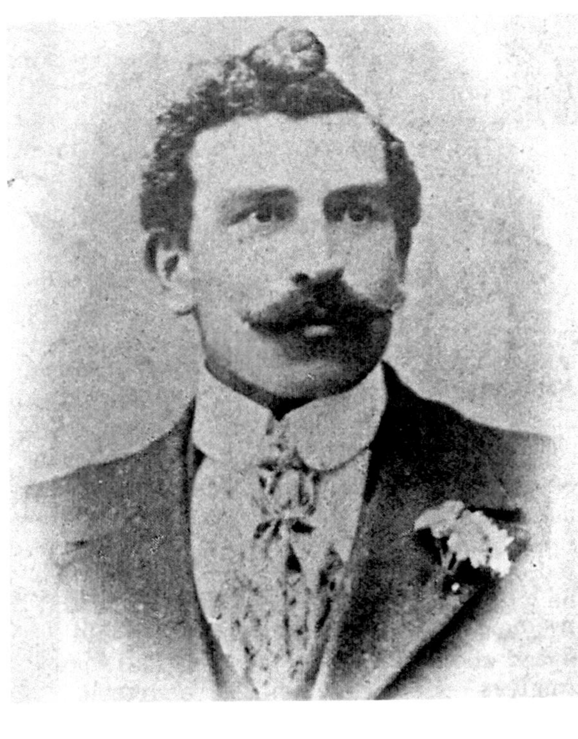

Antonio Poli. *AIMH Archives*

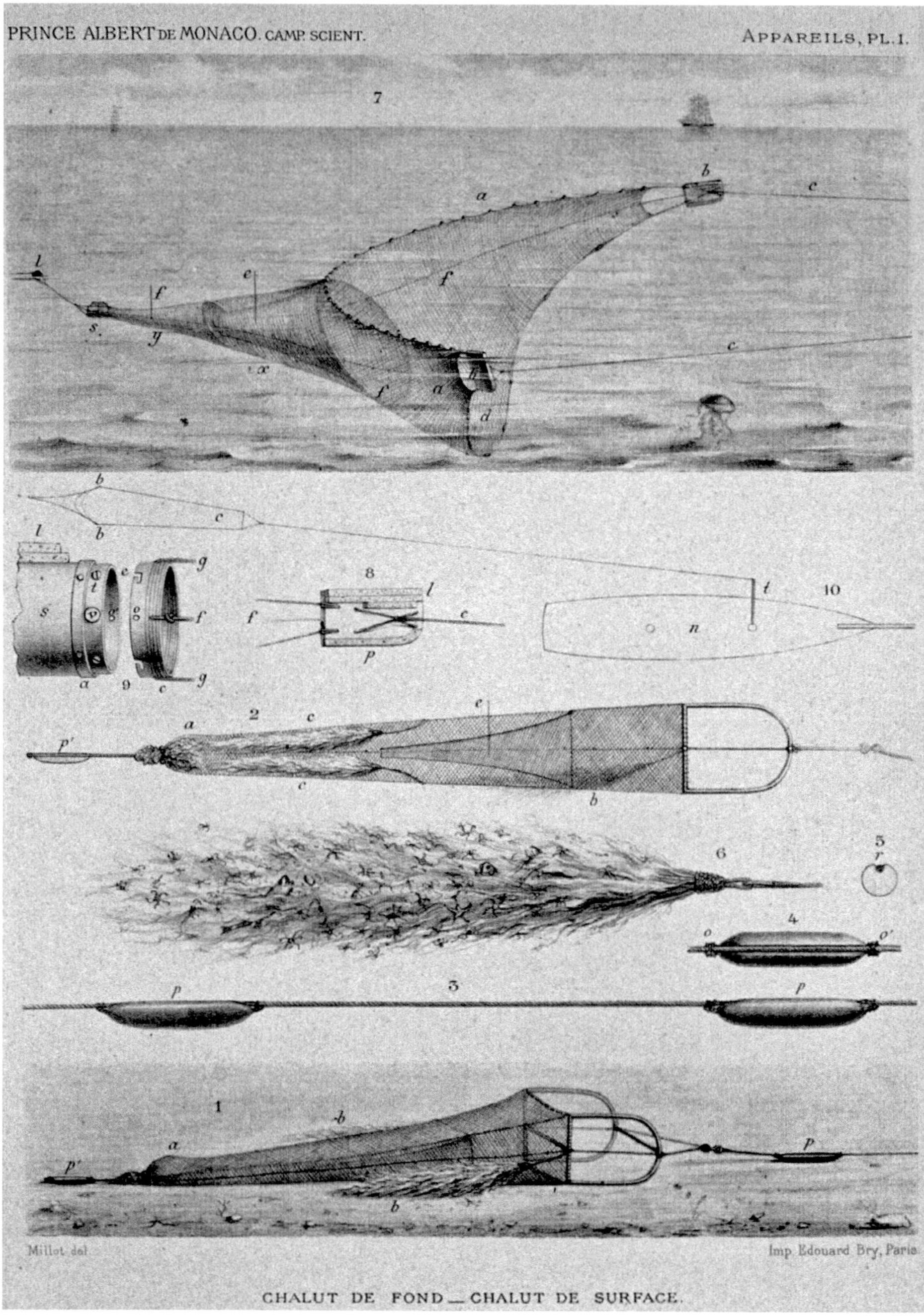

Early Depiction of an Otter Trawl.

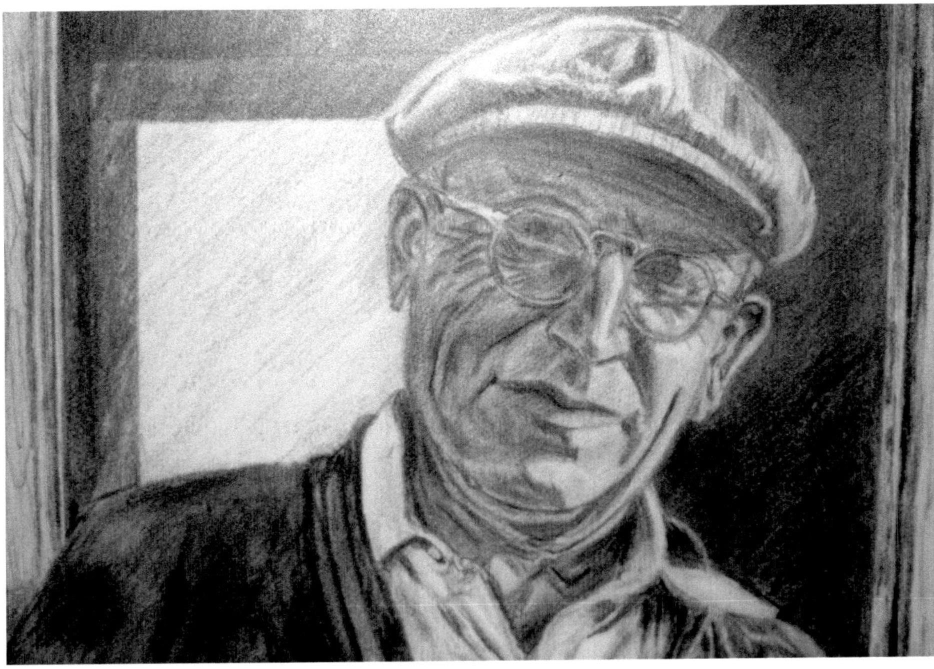

Sketch of Salvatore Tringali. *AIMH Archives*

by 1917, and Florida's annual production of shrimp was over 10 million pounds, worth about $200,000. Roughly 950,000 pounds of shrimp passed through Fernandina's shrimp houses and canneries.

In a way, Amelia Island's shrimping success ultimately contributed to the end of its place as the epicenter of modern shrimping. As the industry boomed in Fernandina, the docks became crowded and the shrimp off of Amelia Island's coast began to be fished out. Shrimpers followed the catch down to Cape Canaveral. The distance made a base of operations in Fernandina inconvenient, and many shrimpers made the move south to St Augustine. Around 1922 Mike Salvador began renting land in St Augustine at 158 King Street, along the St. Sebastian River. He built docks and a fish house there, and was soon joined by the Versaggis, Poli, Carinhas, Tringali, DeGrande, and Felice Golino. The families maintained residences in both Fernandina and St. Augustine, and continued to do business in both cities. Fernandina ceased to be the hub of the shrimping industry, but it could still count itself among the key ports on Florida's East Coast during shrimping season.

Mike Salvador died suddenly in 1924, followed by Salvatore Versaggi six months later. Mike and Salvatore left their families ten boats each. Still, Salvador's widow was left with four small children to provide for, while Versaggi's widow Vincenza had seven. Tony Versaggi's wife Concetta had passed away in 1918, joined by Joseph Tringali's wife Virginia

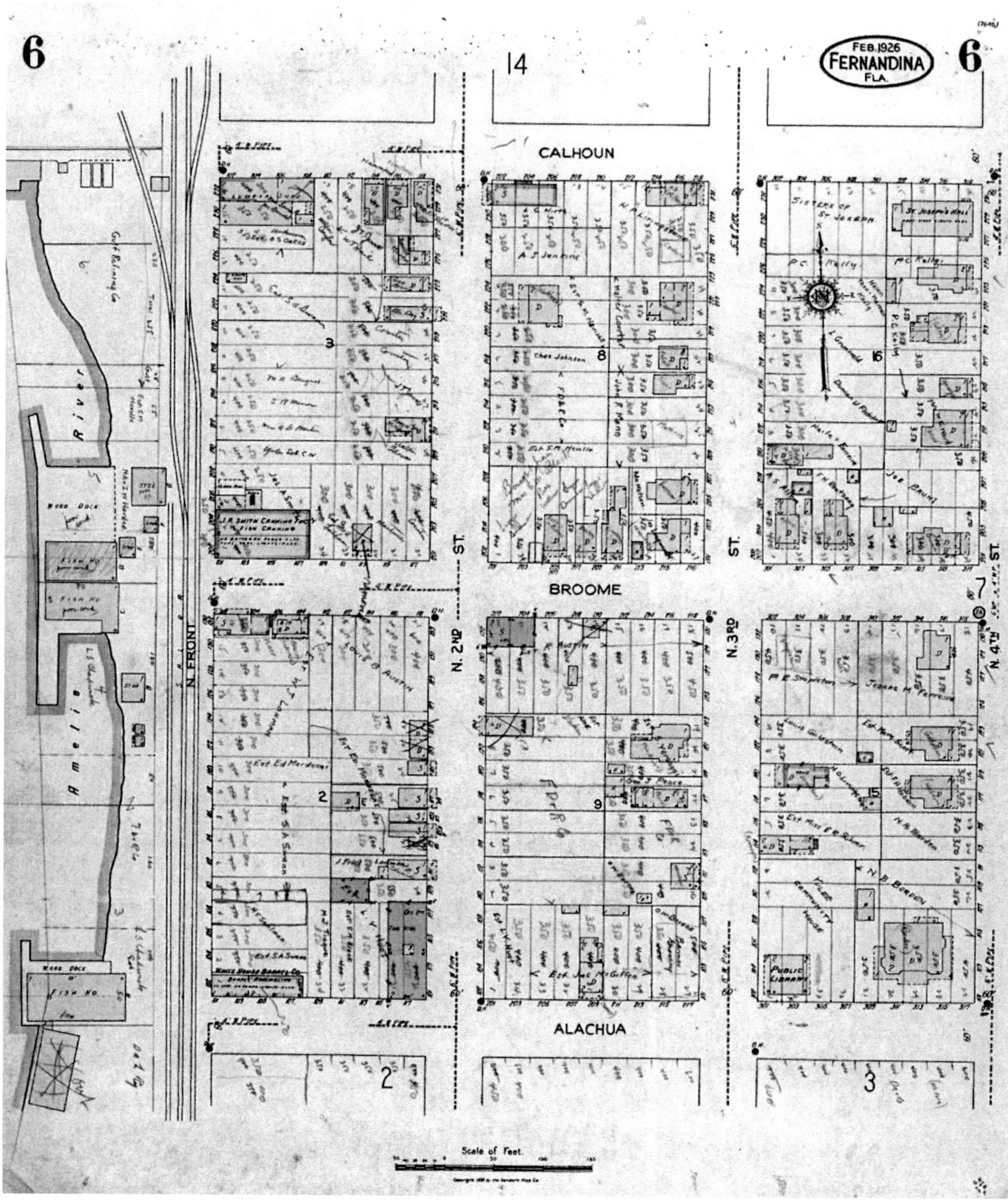

This map of Downtown Fernandina, made for insurance purposes in 1926, shows several of the fish houses on the waterfront at that time, just north of Centre Street. *AIMH Archives*

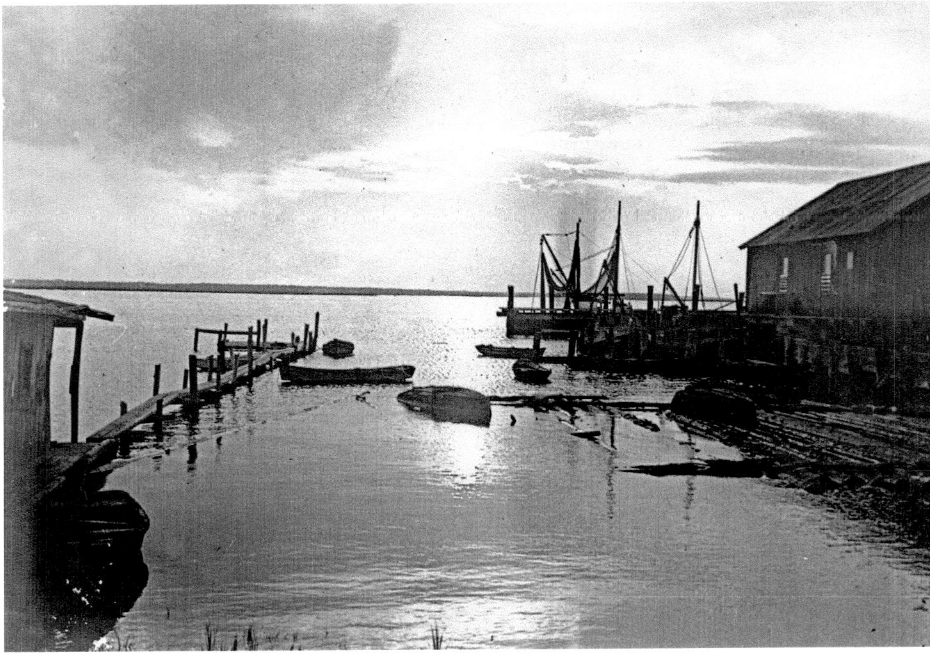

Fernandina's River Front. *AIMH Archives*

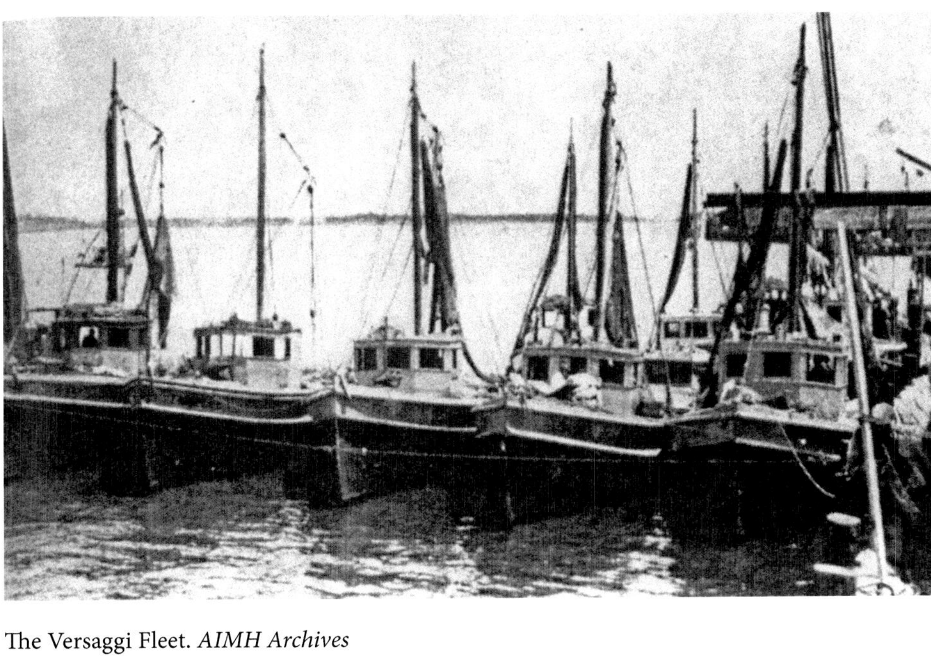

The Versaggi Fleet. *AIMH Archives*

(younger sister of Salvatore and Tony Versaggi) in 1920, leaving both men widowers, each with six children in their care. The extended family did their best to provide for these children. Many were raised by relatives, though some had to spend part of their youth in orphanages. The Versaggi, Salvador, Tringali, Litrico, and Poli families for the most part remained extremely close and their various business holdings became even more integrated.

Antonio Versaggi learned to read and write English after his older brother's death and along with Antonio Poli, took on a larger role in running the families business interests until their nephews came of age and could become more involved. Tony Versaggi operated six boats of his own, and managed the Salvador fish house. His friend John Santos Carinhas took over the marketing. Shortly afterwards, Carinhas proposed marriage to the then sixteen-year-old Frances Tringali. Many relatives opposed the coupling because John was more than twice her age, but Tony Versaggi gave his blessing because he believed Carinhas was a good man and would be a solid provider. John and Francis married in September of 1925. The marriage lasted until John's death in 1975, and produced five children.

The shrimping industry as a whole continued to thrive through the 1920s, until bottoming out at the start of the Great Depression. When the banks failed, shrimpers found it hard to afford to operate their boats and many lost everything. This break in commercial activity gave many shrimp grounds which were in danger of being fished out time to recover. By the mid-1930s the shrimp population was restored, and catches remained plentiful until another rut came in the 1940s.

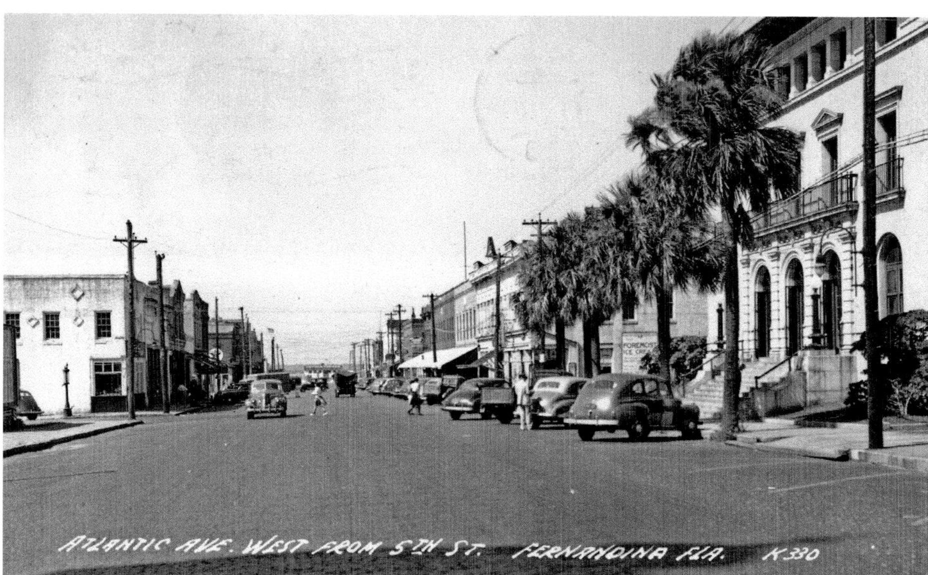

Centre Street, Downtown Fernandina in 1945. *AIMH Archives*

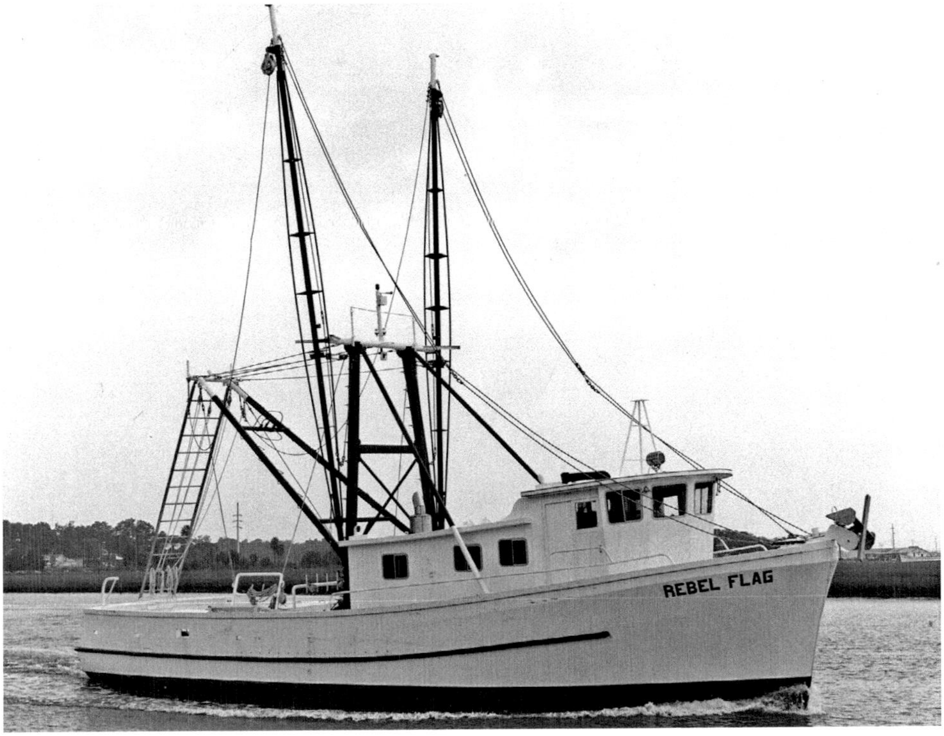

Local trawler the "Rebel Flag." *AIMH Archives*

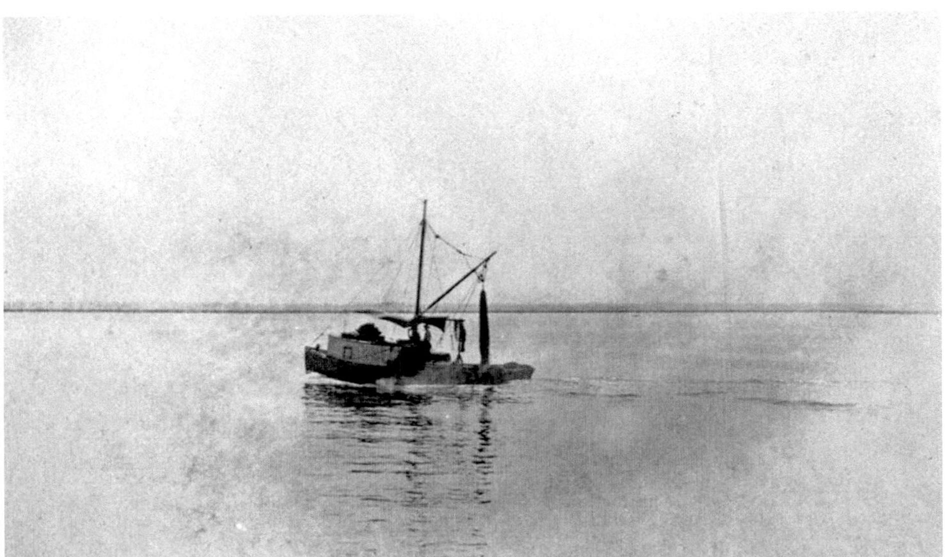

A lone Trawler fishing near Fernandina. *AIMH Archives Courtesy of Billy Burbank III*

As local waters were fished out, shrimpers started traveling further and further away in the search of new shrimping grounds. The trips became longer, but shrimping continued to be the livelihood of many local families. In the mid-1930s Fernandina dealers received an average price of 15 cents per pound of shrimp. During the season, around 37,500 pounds of shrimp were shipped north by truck each month. In 1936 the local Railroad Agency reported shipping 6,935 barrels by train between June and September. The average shrimper could earn $75 for a week's work (sometimes up to $100 on an especially productive outing). However, it is important to keep in mind that the work, along with the earnings, only came during the season.

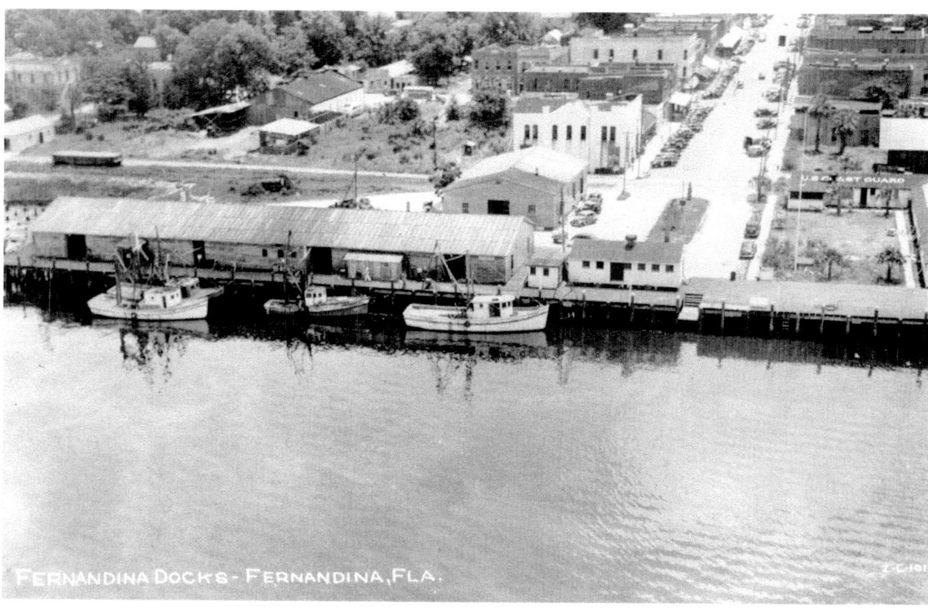

Downtown Fernandina in 1943. *AIMH Archives*

3

Branching Out: Pink Gold

In the mid-1930s, Mike Salvador's sons Felix and John were ready to enter into the family business. They repainted the Salvador fish house to read S. Salvador and Sons. Felix ran the shrimp boats and John handled marketing. Domenica Salvador, now settled into the role of matriarch, remained a strong influence over her sons and the business. John Santos Carinhas went into business for himself in 1928, adding seven boats to the one he had already owned. Carinhas had earned a reputation not only as a seasoned fisherman, but also an excellent marine mechanic. Manufacturers valued his input, giving him experimental engines to outfit his boats with.

In 1931 Carinhas moved his boats to Young's Island, South Carolina, Honduras, then Texas, in search of more plentiful shrimping grounds. Finally settling in Morgan City, Louisiana, in 1938 he discovered the most lucrative concentration of shrimp up to that point. John convinced many of his friends and family (who fifteen years earlier had moved from Fernandina to St. Augustine) to relocate once again to Morgan City. By the late 1930s, St. Austine's shrimping industry was in decline. The sons of Salvatore Versaggi, John, Virgil, Dominic, and Manuel had to borrow money to put gas in their boats for the journey. Leaving their uncle, Antonio "Tony" Versaggi in charge of their operation in St. Augustine, the brothers intrepidly headed for the Gulf.

Soon, Morgan City was in a full blown boom. National publications like *Reader's Digest*, *Colliers*, and the *Saturday Evening Post* ran stories on the thriving industry. *Southern Fisherman* claimed, "…there were more Florida people than local people in the shrimp industry." For most shrimp boats at that time, coming back to port with forty to fifty barrels of shrimp was considered a good run. John Santos Carinhas reported one of his ships returning with 156 barrels in one trip.

During World War II, demand for shrimp rose through the roof. Jumbo shrimp was selling for twenty cents a pound. The Versaggi Shrimp Company thrived in Morgan City. The Veraggis started a retail shrimp business and opened an ice plant. Their fleet of shrimp boats went from six active boats to twenty five. With Virgil serving the Coast Guard, and Dominic in the Navy, Uncle Tony came to assist the brothers with running their thriving fish market.

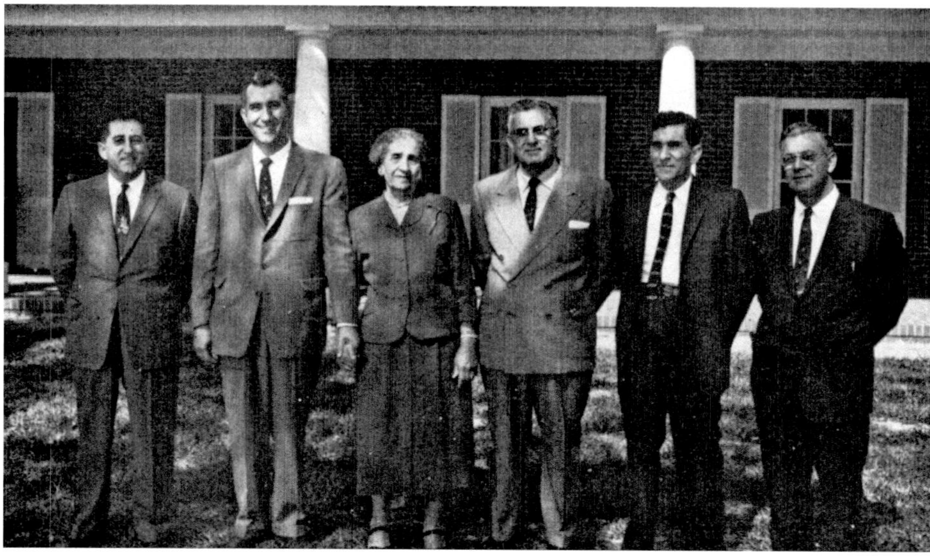

The Versaggi brothers with their mother Vincenza. (From left to right): Virgil, Dominic, Mrs. Versaggi, John, Joe, and Manuel.

Dominic and Virgil become well acquainted with the most up-to-date nautical technology, radar, and sonar equipment during their time in the military. After the war, they brought that knowledge back and applied it to the boats in the Versaggi fleet. At the same time, machines were being created that simplified the procedures for cleaning and packing shrimp. Business was good; Manuel Versaggi recorded one occasion when a storm forced all boats in the area into port and kept them there for five days. Versaggi bought every pound of shrimp in Morgan City and made $8,000 reselling it.[3]

In 1950, the Salvador family was responsible for another key moment in the history of the shrimping industry. For many years shrimp had been turning up in bait nets hung between bridges in the Florida Keys, and Key West fisherman had found large shrimp in the stomachs of fish they caught. The Marine Laboratory at the University of Miami had previously published predictions of sizeable shrimp populations southwest of Florida, and many shrimpers had trawled the Keys trying to find them, but to no avail.

Brothers John and Felix Salvador (the sons of Sollecito), who were then working out of St. Augustine, decided to travel to Key West to take a look for themselves. Joined by veteran shrimper Bluff Peterson, the brothers borrowed a battered old fishing boat from a local seafood firm and headed south. They installed a try-net and embarked on a scouting expedition of the Dry Tortugas. After two days without any success, the boat was ready to turn back to port. Felix Salvador relayed what happened next to National Geographic Magazine. "We were heading back for Key West, but as one last chance, we let the trawl down one more time, even though the sun had set. When we hauled it in, we found it

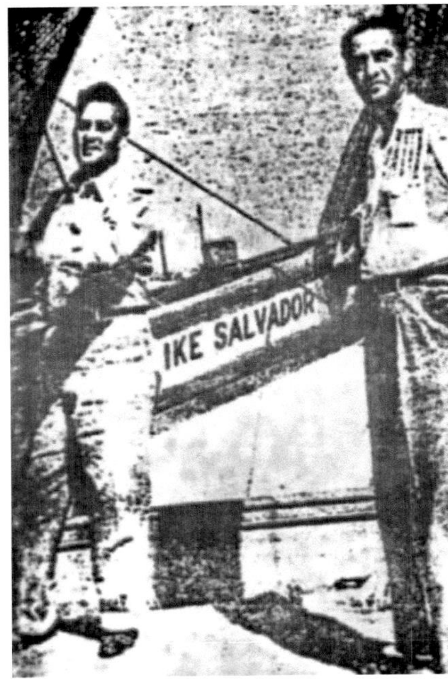

John and Felix Salvador.

bulging with beautiful big pink shrimp. We found the secret: those shrimp came out of the mud only at night!"

The Salvador's success would not stay a secret for long. They returned to port with their nets overflowing, and sent their entire fleet south. Word of the discovery swept up both coasts and shrimpers from the gulf all the way to North Carolina converged on the Dry Tortugas. The Key West grounds became a bonanza of what the shrimpers called "pink gold," and Key West became Florida's chief shrimping port seemingly overnight.

By 1957 shrimp had become the most valuable seafood item in the United States, bringing shrimpers a national annual income of around $65,000,000 (tuna, the next closest earner came in at around $48,000,000). White shrimp (until then the most common at most market places) was put on the back burner. Brown and pink shrimp now accounted for ninety percent of the Florida catch.

Tony Versaggi was sixty years old when Morgan City began its boom. He joined his nephews there to help them look after their fish market. He appeared on a 1945 cover of Liberty Magazine smoking a corncob pipe and holding up two very large shrimp, with a caption that read "not all shrimp come little." Shortly after World War II, Tony suffered an illness so serious he was pronounced dead. The nurse had already filled out the death certificate, but somehow he rallied and went on to live another twenty years. After a long convalescence with his daughter Nellie in New Jersey, Tony returned to Saint Augustine to be the caretaker of the Carinhas home. Tony Versaggi spent his last years surrounded by nieces, nephews, and their children. He passed away in 1970, at ninety-two years old.

In 2000 Sollecito "Mike" Salvador's achievements were recognized when the state of Florida named him a Great Floridian. His home at 20 S. 4th Street still stands, marked by a plaque which denotes his indelible impact on the modern shrimping industry. He is buried in Bosque Bello Cemetery in Fernandina, surrounded by his family and fellow shrimping pioneers: his wife Domenica, Salvatore and Tony Versaggi, their wives Vincenzina and Concetta, Salvatore, Virginia, and Francesca Tringali, and many others.

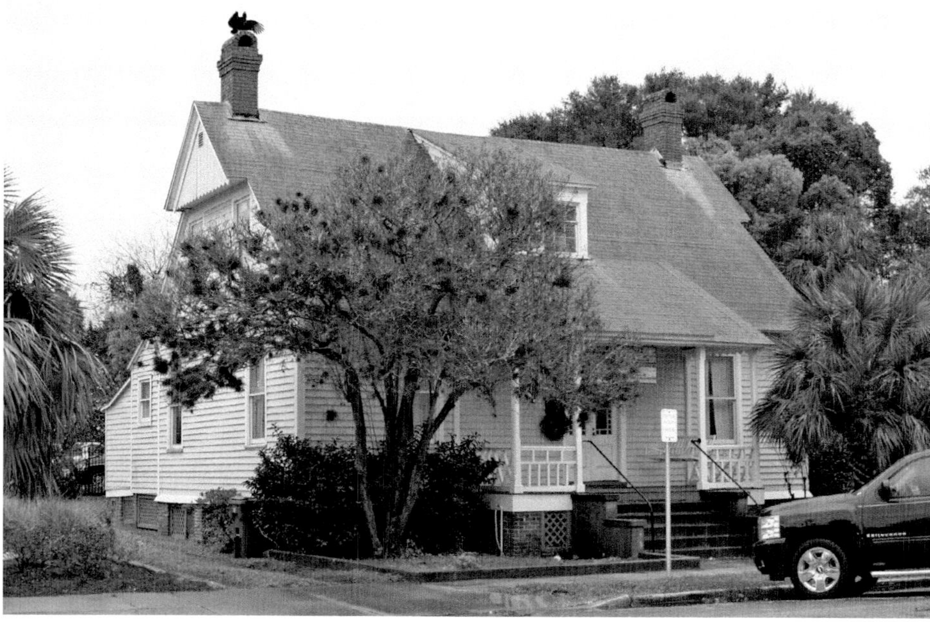

The former home of Mike Salvador, photographed in 2015. *Photo taken by Author*

4

Salt Life

THE PEOPLE WHO MAKE their living shrimping have a unique and rich culture all their own. Shrimping is a seasonal endeavor, with the main season in the area running roughly from June to January. Brown shrimp are caught from June to August, and white shrimp are harvested starting in September. Shrimpers sometimes have to rise from their beds before dawn, and other times they stay out fishing all night. Some species of shrimp (like pink and brown shrimp) are nocturnal, and are best caught at night while they feed.

Depending on the time of year and the location of the catch, they sometimes have to travel far from home and stay gone for several days following shrimp. Spending so much time away from home can put a strain on family life. Many shrimpers pursue other sources of income during the off season. They often fish for other types of seafood, and pursue various other jobs to put food on the table.

For many years Amelia Island also had a thriving menhaden (also known as pogy) fishing industry. Pogy are small fish that swim in schools, which are commonly used to make fertilizer and pet foods. Whereas most shrimp boats had crews of only three or four men, pogy boats needed much larger crews to operate. Pogy boats could have up to thirty men working on board. A pogy boat carries two smaller boats called "purse boats." When a school of menhaden are spotted, the purse boats are loaded into the water. They circle the fish with a large net called a "purse seine" and load the catch onto the main boat.

Fernandina had its own pogy plant called Nassau Fertilizer on the Amelia River, just south of Old Town. The original company began as the Seminole Oil and Fertilizer Company in 1911. T. J. Corbett joined the company in 1926, followed by his brother Exteen in 1949. The Corbetts acquired a majority share of the company, and the company remains in the family to this day. Like most of the other fishing families in the area, the Corbetts also spent part of the year shrimping.

Shrimpers who own their own boat are in business for themselves, but the crewmen who work on someone else's boat are usually paid based on a percentage of what they catch (often around ten percent).To give some estimation of a boat owner's expenses in rough numbers, a 50 to 70 foot fiberglass boat can easily go for $150,000 to $200,000 (not including electronics – which can cost up to $30,000 separately). Add in $10,000 for rigging,

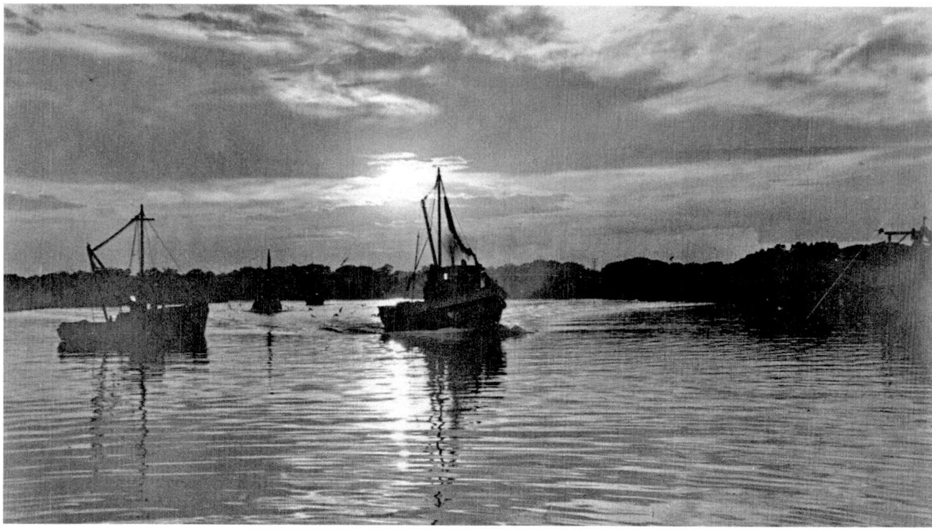

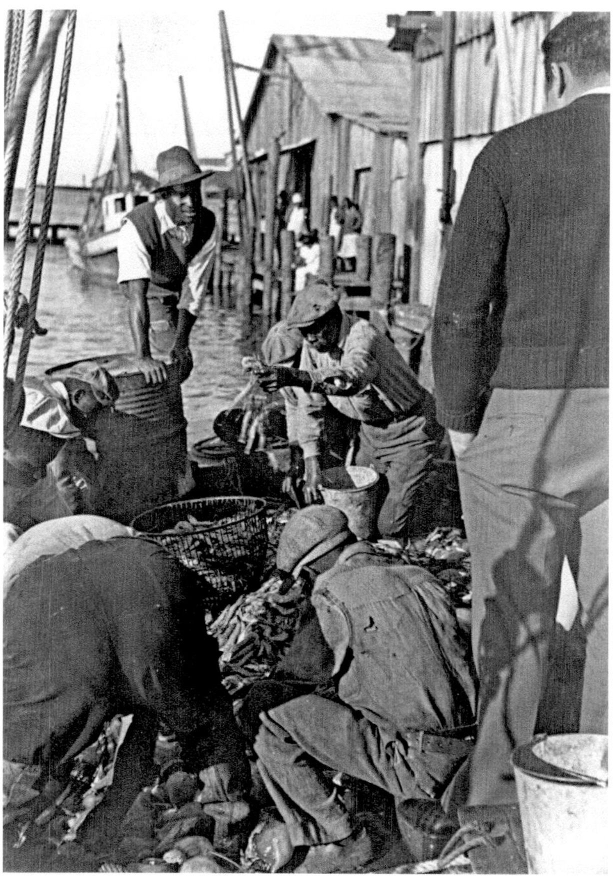

Above: Shrimp boats returning to port. *AIMH Archives*

Left: Sorting Shrimp on the dock at Fernandina. *AIMH Archives*

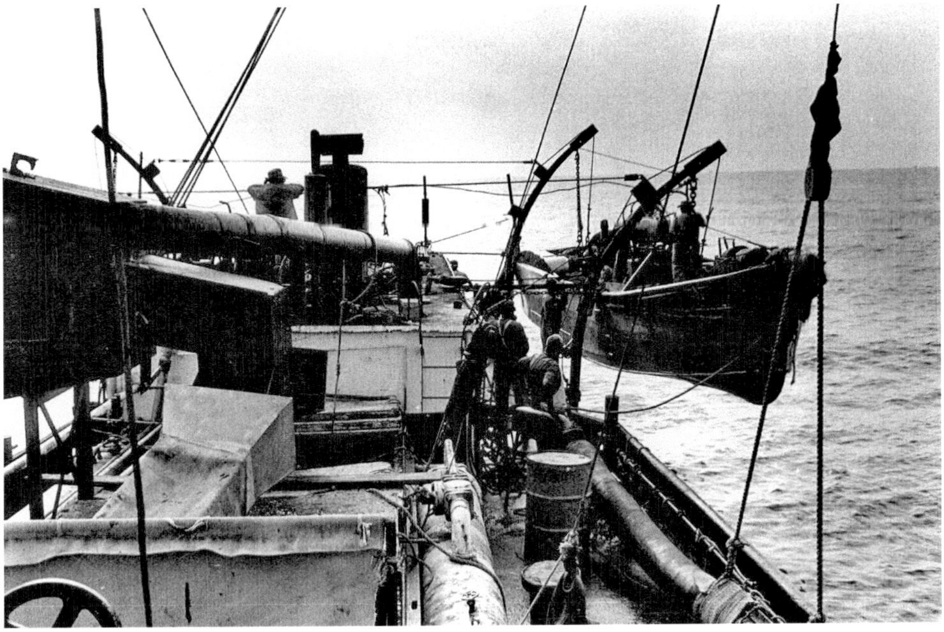

Purse boats deployed into the water. *AIMH Archives courtesy of Billy Burbank III*

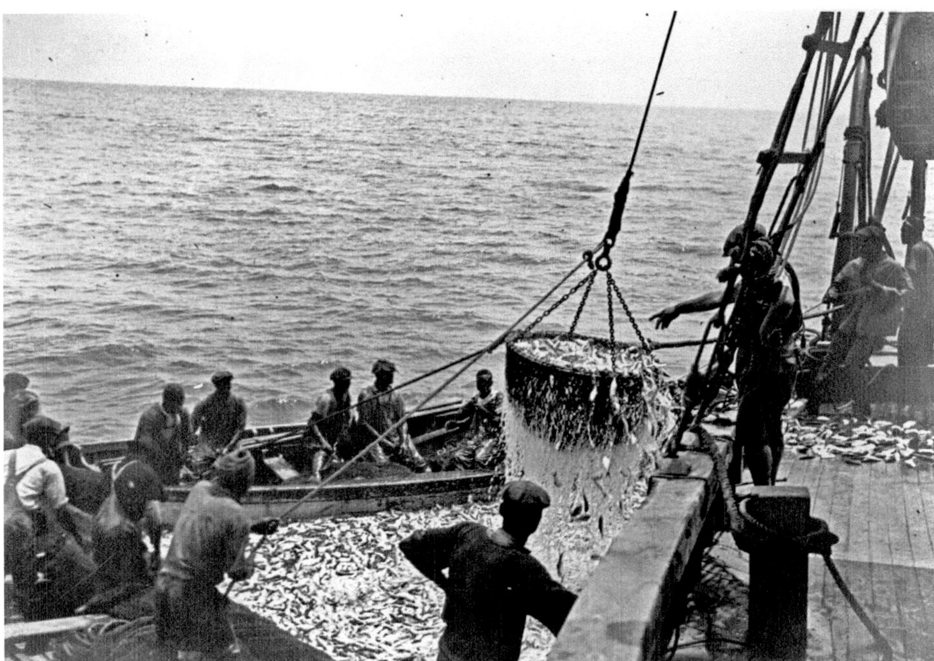

Pogy Fisherman loading the catch. *AIMH Archives courtesy of Billy Burbank III*

ropes, cables, etc. (conservatively), $800 per pair of trawl doors, and about $1600 per net. Include the always trending up costs of insurance and diesel gasoline to paint a vague picture of the costs of owning a commercial shrimp trawler.

Like other seafarers, shrimpers can be extremely superstitious. Some of these beliefs are almost universal among mariners. For example, many captains don't allow whistling on their boat. In the old days, many crews were reluctant to bring a woman on board for fear the act would bring misfortune – a pregnant woman would spell utter disaster.

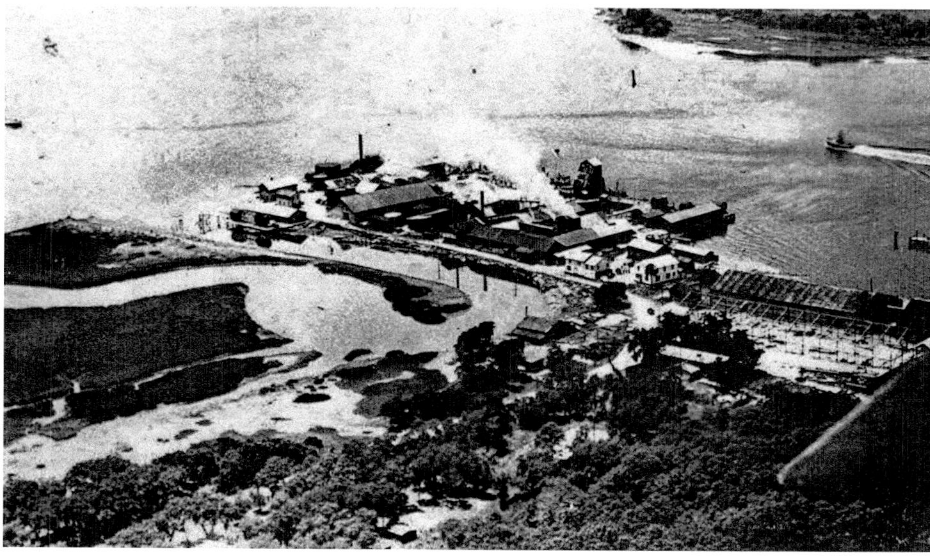

Nassau Fertilizer, Fernandina's Pogy Plant. *AIMH Archives*

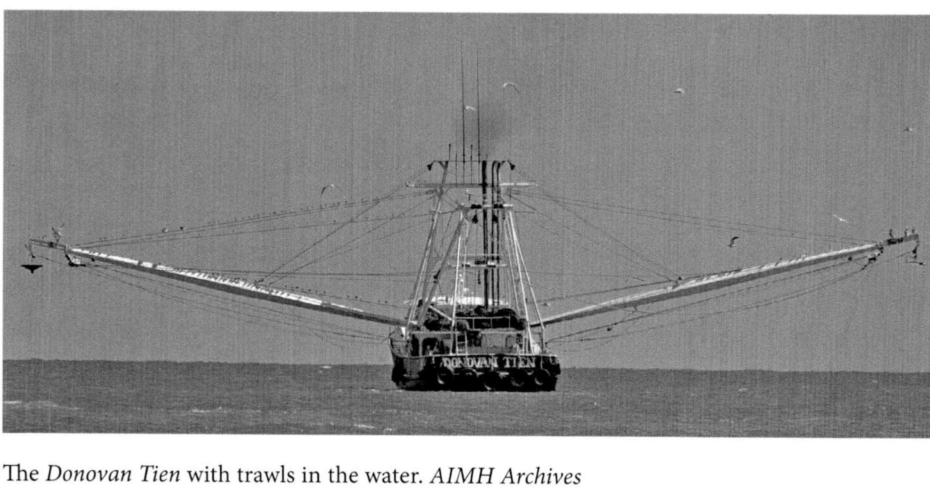

The *Donovan Tien* with trawls in the water. *AIMH Archives*

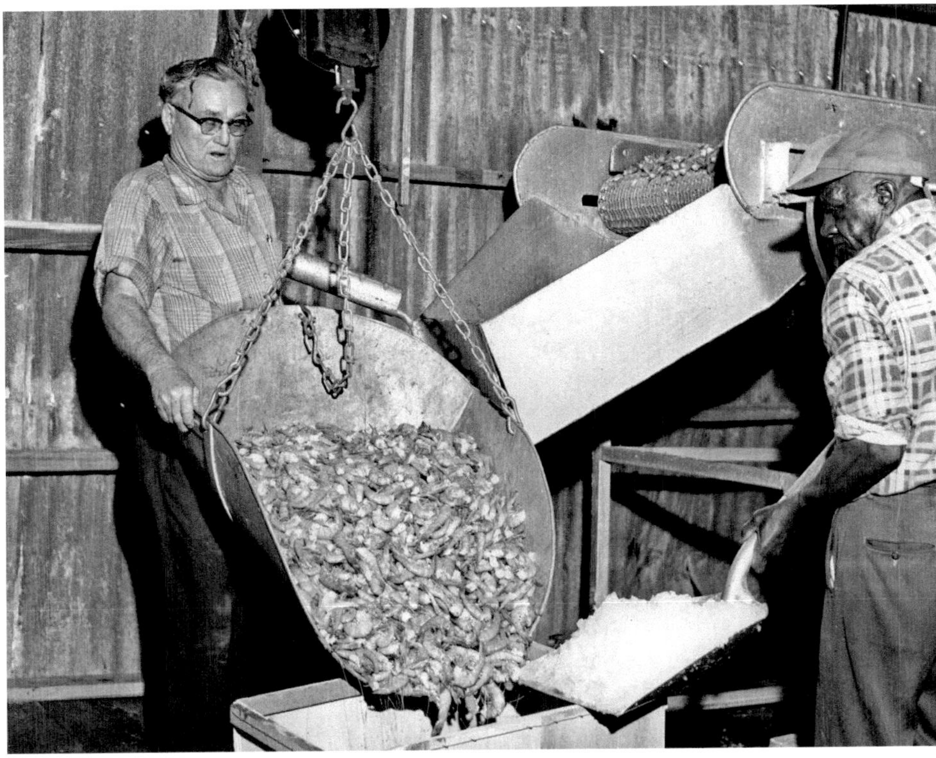

Sloan Peterson and Frank "Porkchop" Timmons packing shrimp. *AIMH Archives*

There are some idiosyncrasies that seem unique to the region of northeast Florida, whose origins have faded into folklore. For example, shrimpers throughout the area have an absolute abhorrence of alligators, and it is a cardinal sin to say the word, or any derivative (like gator) on a shrimp boat. When taking off a hatch cover, one always has to set it down face up, just as it sits when in use. Some captains do everything in their power to avoid starting a trip on a Friday because they associate the day with malevolent fortune. Another interesting facet of the culture involves the singular and sometimes peculiar nicknames bestowed on shrimpers. Examples of colorful sobriquets belonging to local shrimpers include: Cap'n Eat Em Up, Albatross, Cape Canaveral Bill, Seven Finger Brown, Blue Nose Charlie, and Jitterbug Joe.

Shrimping can be a tough and strenuous life. A shrimper's hands are densely calloused, from the top of the fingertip to where the wrist meets the palm. It is rugged, physical work, and like any profession on the water it can be extremely dangerous. The first shrimp boats ran on primitive fuel, which was highly flammable and prone to ignite with the slightest spark. Early shrimpers had to use manpower to maneuver the ropes and pulleys used for hoisting nets. As technology progressed, more and more mechanical equipment found its way onboard and provided its own hazards.

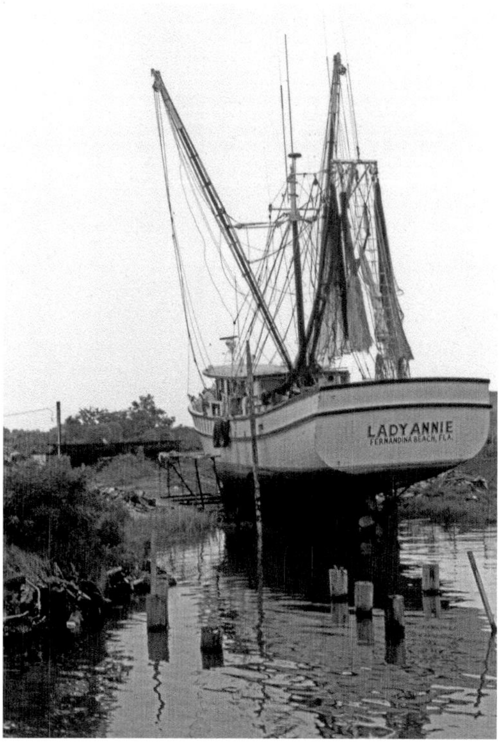

Local shrimp boat the *Lady Annie* in dry dock for repairs. State Archives of Florida. *From the Collection of Ormond Loomis*

Shrimping provides an income not only for captains and crews, but for several complementary professions that go hand in hand with it. Shrimpers require the services of boat builders, mechanics, and net makers. 'Headers' peel, and if needed, pop the heads off shrimp after they are unloaded from the boats, before they are packed to be shipped. Heading was a source of extra income, particularly for the young, the old, and people in between jobs. Headers were paid based on productivity, and in the 1950s could earn up to two or three cents per pound. Distributing shrimp is a field unto itself. Workers are needed to load the boats with ice and to load and unload shrimp when it's shipped. Truck drivers are required to transport it, and retailers to sell it.

There is a strong family dynamic to shrimping. Like many other trades, it tends to span several generations in a family. The only way to learn is through watching and doing, and many shrimpers begin their apprenticeship at a young age. Along with experience, boats and shrimping equipment are often a big part of the inheritance shrimpers pass down to their children. The boats crafted at Fernandina were of impressive quality and some are still at sea in some capacity to this day.

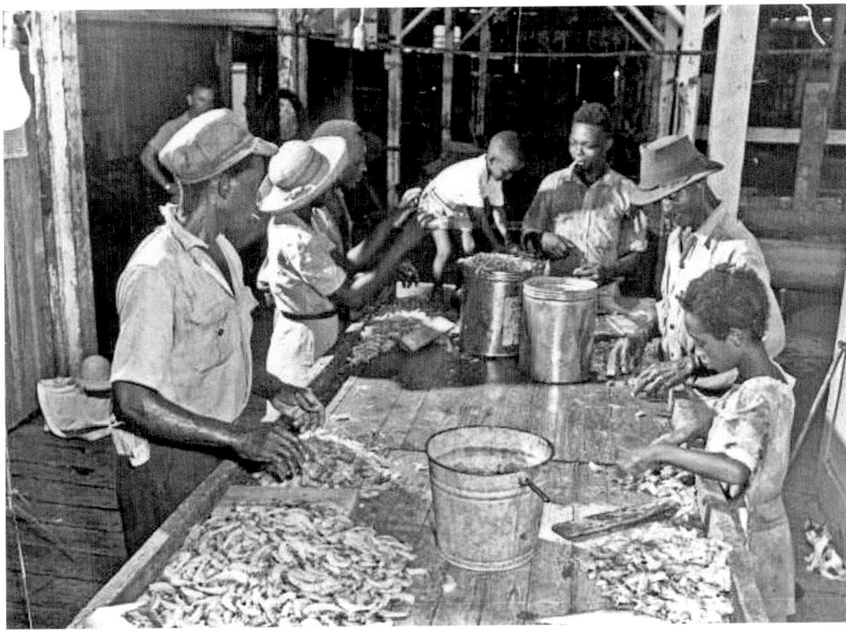

Headers at work. *AIMH Archives*

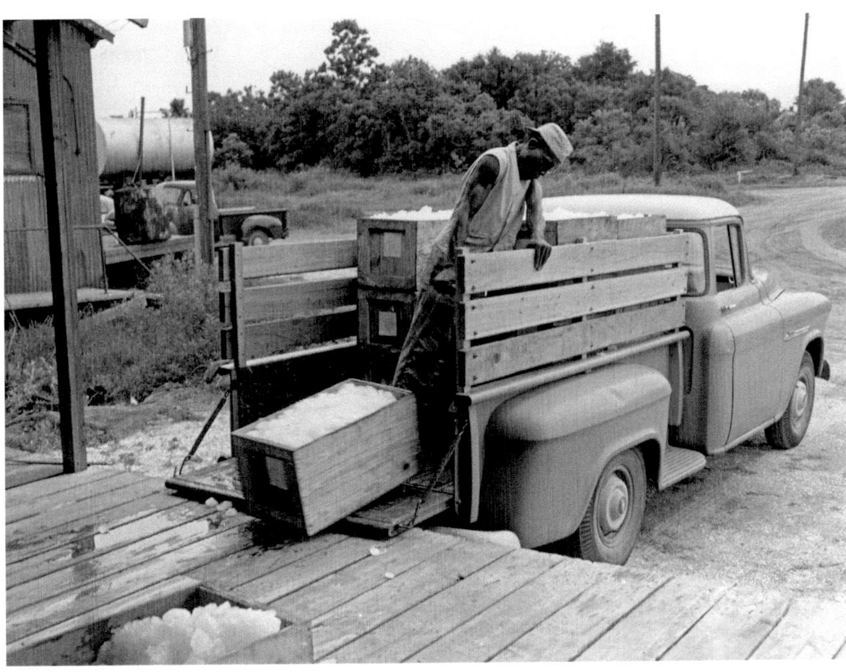

Loading shrimp at Fernandina for packaging in Jacksonville—1957. *State Archives of Florida; photo by Karl E. Holland*

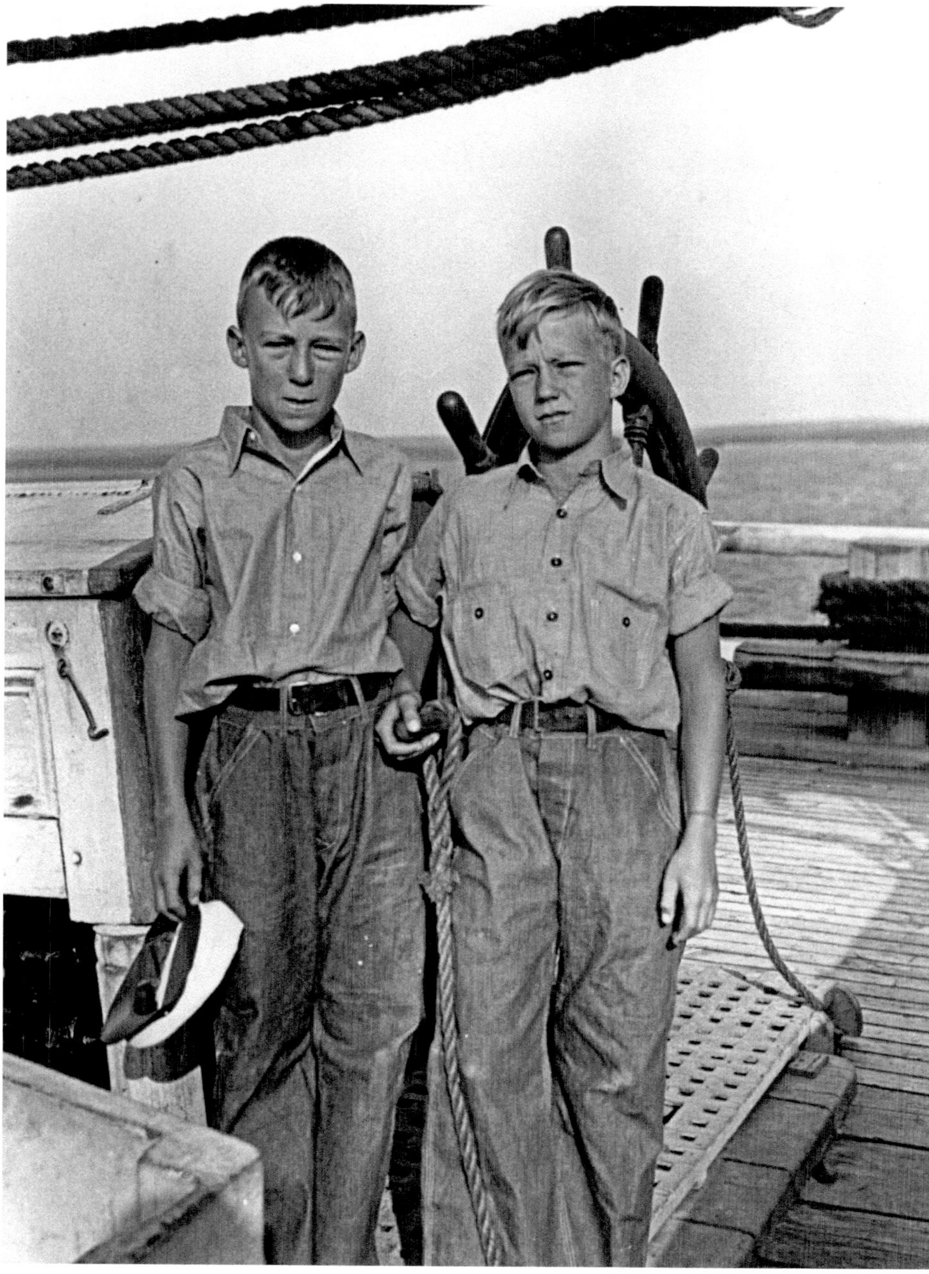

Two young boys on a shrimp trawler. *AIMH Archives*

5

Boatbuilding on Amelia Island

As a coastal city, there has always been a demand for all types of water craft in Fernandina. African American craftsmen like Oscar Danburg, Wizzie Biddle, and Bill Rivers were some of the earliest known boat builders in the area. The shrimping industry boom of the early 1900s led local builders to focus their attentions on shrimp boats. Arguably the most successful of these were two Greek boat builders who chose to practice their beloved trade on Amelia Island.

Michael Nichlos Tiliakos came to America from Greece shortly after the turn of the twentieth century, at just eighteen years old. He settled first in Tarpon Springs, joining the large Greek population there. While visiting Fernandina a few years later he fell in love with a young Irish girl named Clara Franks. Mike relocated to Fernandina in 1912, and began building boats for a man named Mr. Lambros. He married Clara in 1919, and after a few years was able to establish his own business. Over the next fifty years, the boatyard Tiliakos started on the waterfront would produce hundreds of ships, and set the standard for quality boatbuilding in the area.

At its busiest period, Tiliakos' boatyard employed around twenty men and built not only shrimp trawlers, but yachts, barges, pilot boats, and special order crafts. Most shrimp boats ranged from about fifty to eighty feet. Generous and proud, Tiliakos was a popular member of Fernandina's business community. Always willing to help a friend, he was able to weather the Great Depression by bartering land and labor.

Tiliakos owned several shrimp boats of his own, and also built a processing house for processing, packing, and shipping shrimp. Like most of the businesses that revolved around Fernandina's shrimping industry, Tiliakos' boatyard had a strong family presence. Mike's elder son Nick showed an aptitude for business, and managed the family's fleet of shrimp boats. His younger son Johnny was a natural craftsman, who absorbed his father's techniques and became a skilled boat builder in his own right.

One of Tiliakos' greatest assets proved to be a Greek boat builder named Dematrios Nicholas Deonas. Known as "Jimmy," Deonas was born on the Greek Island of Santorini in 1909, to a family of boat builders. At the age of nine he left school and began spending most of his time learning the boat building trade from veteran craftsmen. As a young man,

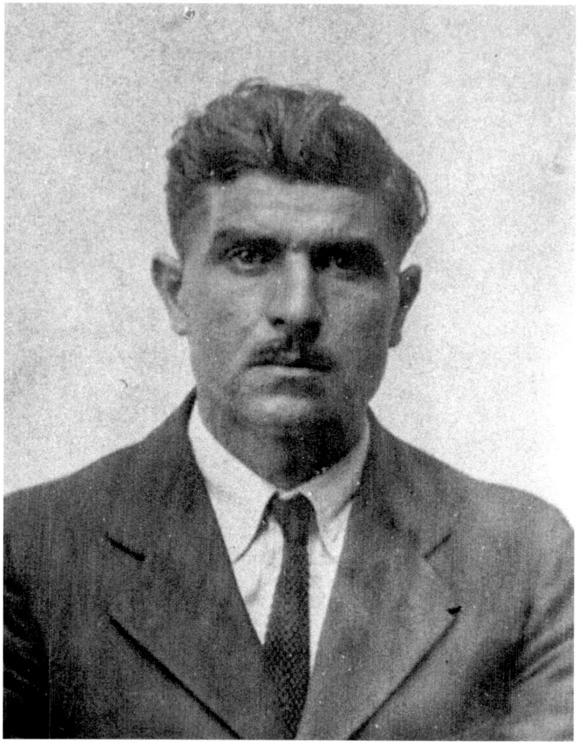

Left: Michael Nichlos Tiliakos. *AIMH Archives courtesy of Nick Deonas*

Below: Tiliakos' Boat Yard on the Amelia River, just south of Ash Street. *AIMH Archives*

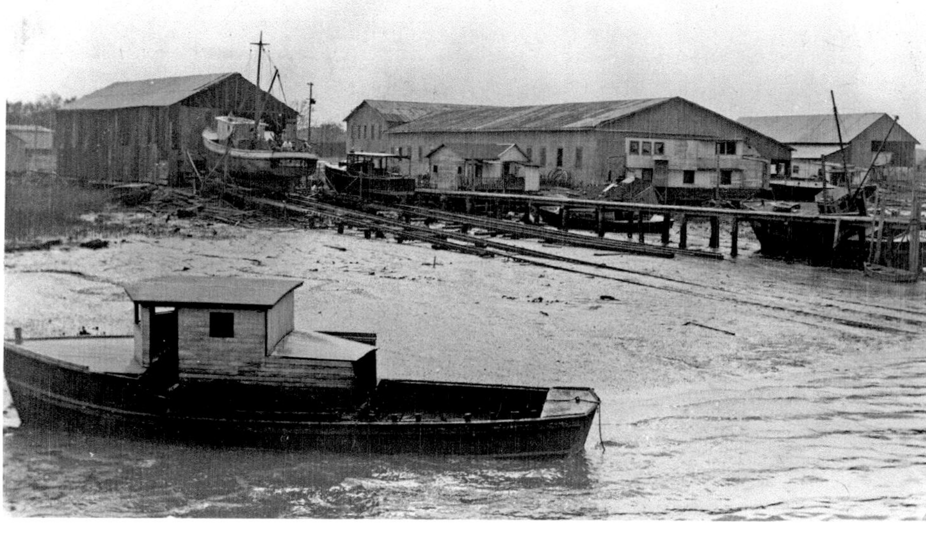

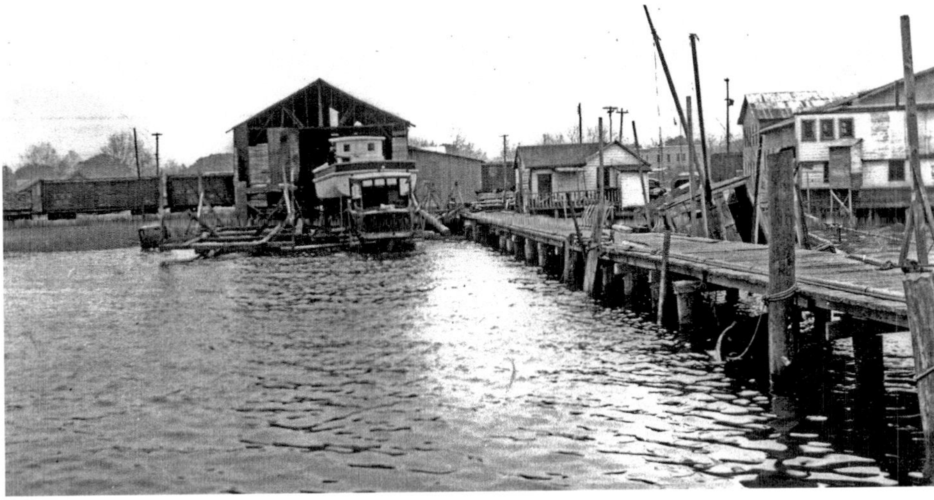

Boat launches were a community event, and people would often come out to watch a new craft being launched. *AIMH Archives courtesy of Nick Deonas*

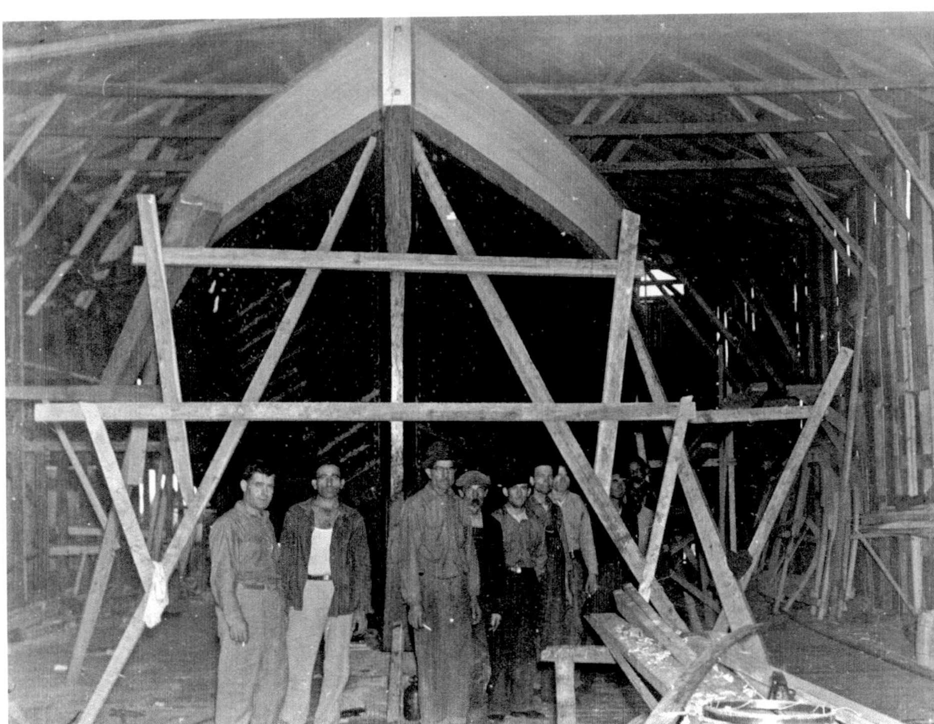

Mike Tiliakos (center) with some of his employees, including Nick Deonas (far left). *AIMH Archives courtesy of Nick Deonas*

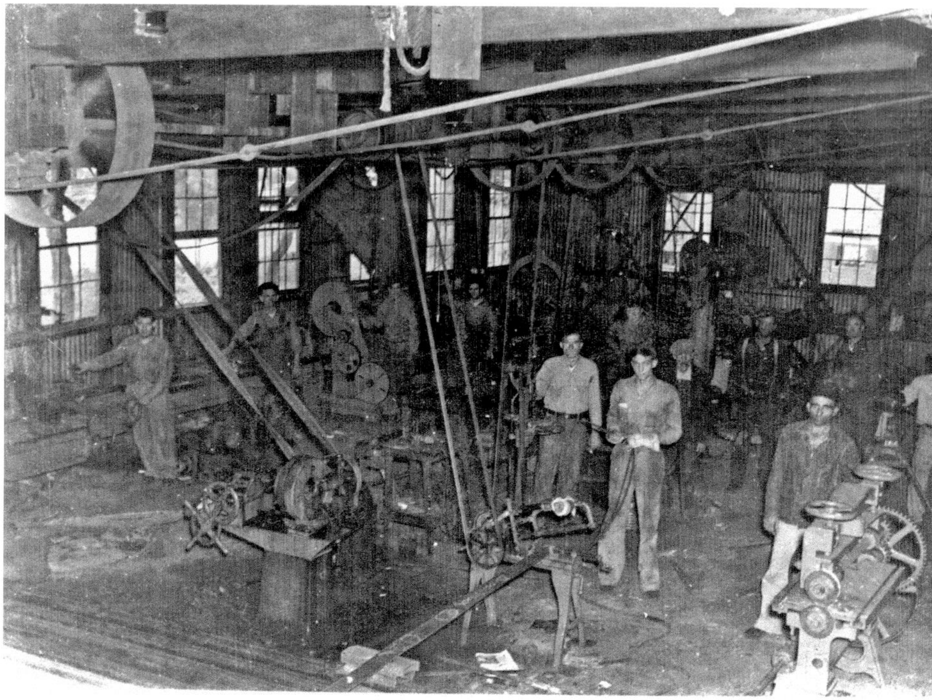

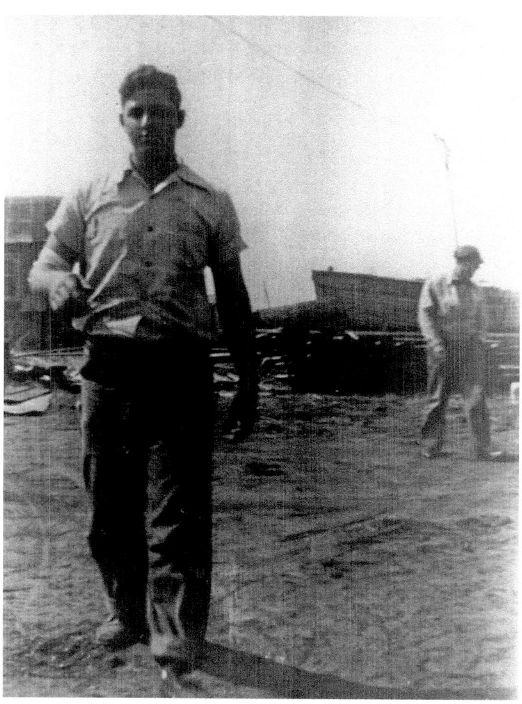

Above: Inside Tiliakos' shop in the early 1940s. *AIMH Archives courtesy of Nick Deonas*

Left: Johnny Tiliako as his father's boatyard. *AIMH Archives courtesy of Nick Deonas*

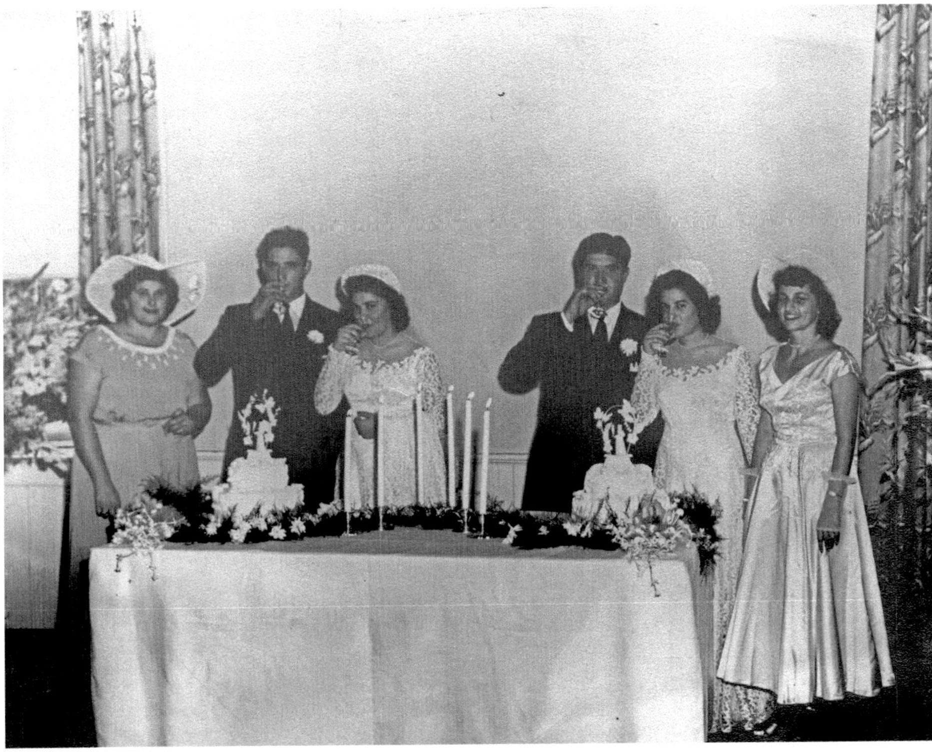

The double wedding of Nick and Johnny Tilakos. From left to right Anna Tiliakos, Johnny Tiliakos, his wife Nellie, Nick Tiliakos, his wife Magdeline, and an unidentified guest. *AIMH Archives*

he joined the Greek Merchant Marine and found himself stranded in New York City at the beginning of World War II. Word of Fernandina's demand for Greek boat builders reached Deonas, and in 1941 he brought his skills and his traditional tools to the Tiliakos boatyard.

Deonas was a master builder with a special gift for nautical mathematics and drawing up blueprints. He revolutionized the shrimping industry by designing what would become known as the "Super Trawler," a large boat with a more powerful engine than its predecessors, which pulled heavy, double rigged nets. This greatly increased production, and the design quickly became the standard in shrimping.

Tiliakos arranged a marriage between Jimmy and his own daughter Anna, which produced four children. With his father-in-law's blessing, Deonas opened his own boatyard just a few miles north on the Amelia River in the mid-1940s. Deonas and Tiliakos continued to enjoy a collaborative relationship, regularly spending time at each other's boat yards. While they built various types of vessels, the Tiliakos/Deonas family specialized in shrimp boats. Orders came from the Carolina's, Georgia, Texas, Louisiana, and Mississippi, and many of the most successful boats in the region hailed from one of the family's shops.

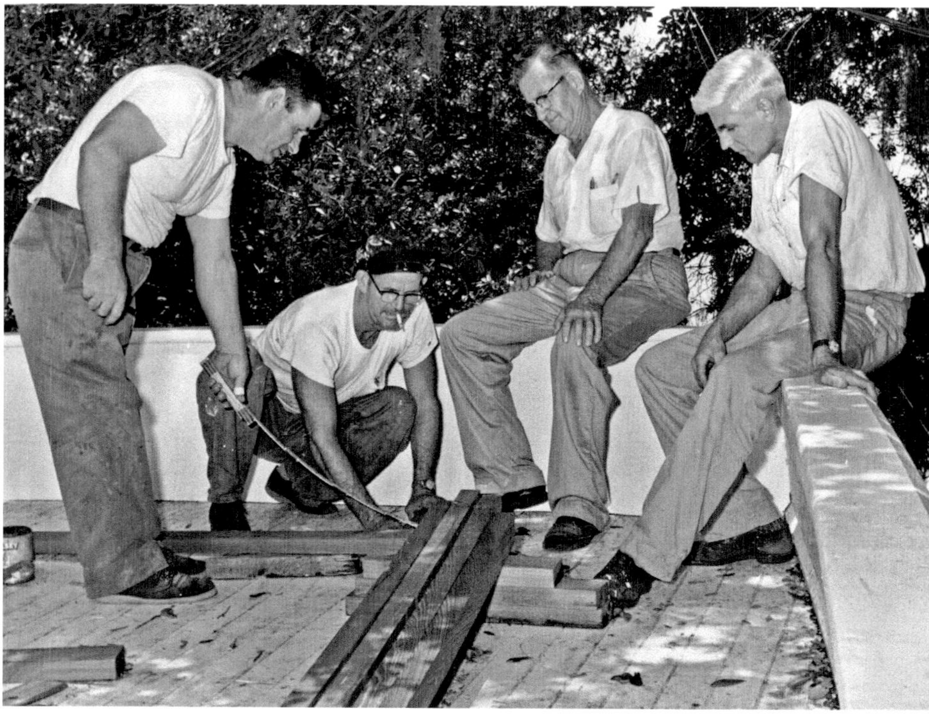

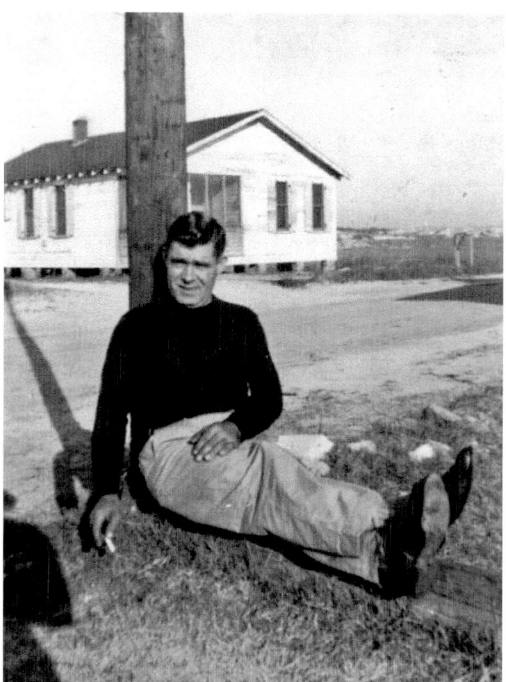

Above: Mike Tiliakos (center) with some of his employees, including Nick Deonas (far left). *AIMH Archives courtesy of Nick Deonas*

Left: Dematrios "Jimmy" Deonas. *AIMH Archives courtesy of Nick Deonas*

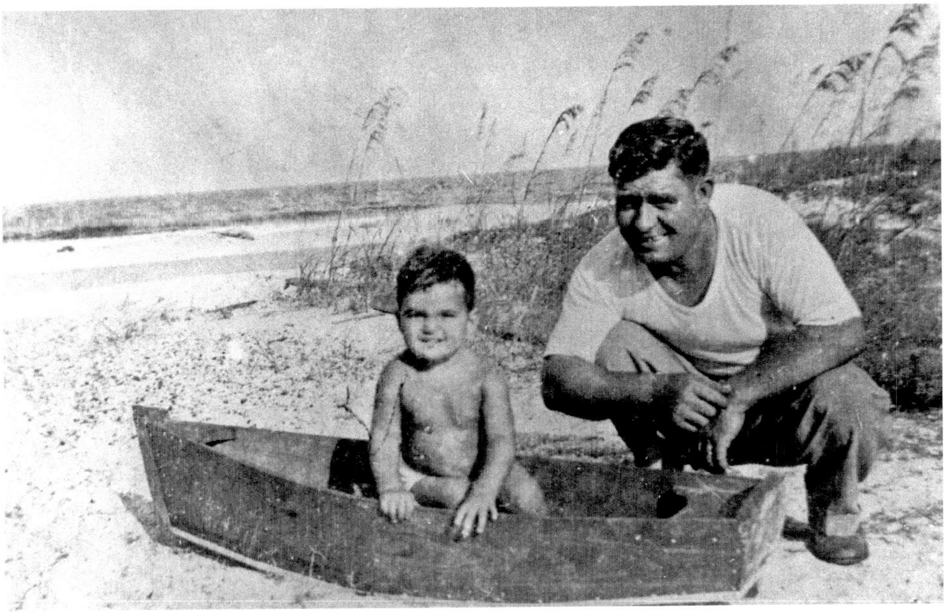

Jimmy Deonas with his son Nick. *AIMH Archives courtesy of Nick Deonas*

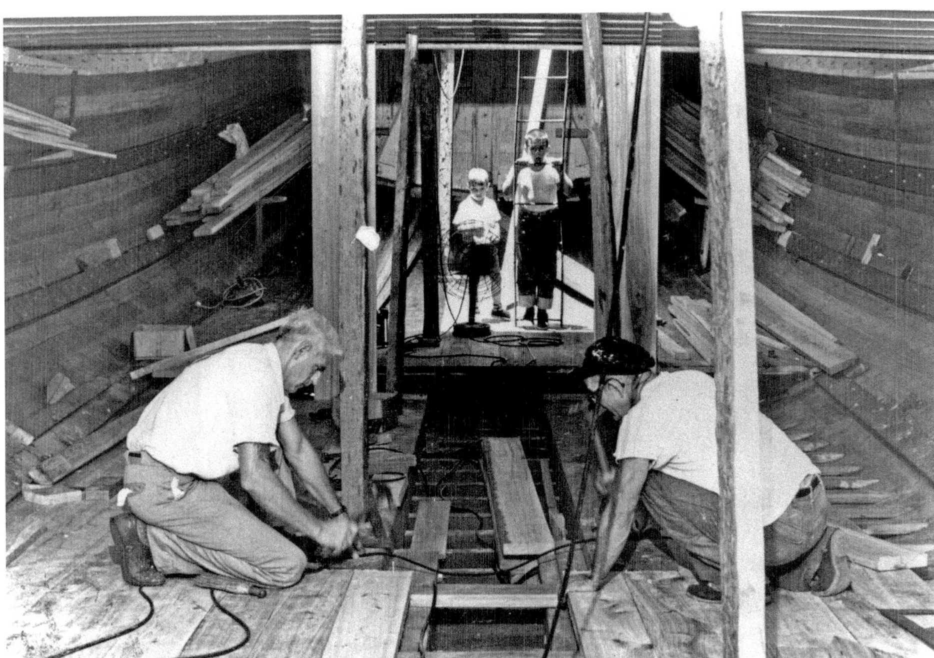

The Tiliakos and Deonas boatyards always had a strong family presence. *AIMH Archives courtesy of Nick Deonas*

Nick Deonas (son of Jimmy and grandson of Mike) describes the process this way:

> Mike and Jimmy built boats the same way, the old world way. Each boat had three natural crook knees of live oak placed in the stern for support. They used the old process of boiling 2 X 4s in clear water and bending them while they were steaming hot into the shape needed. Once the wood cooled the shape was set and placed in the boat, these were the oak ribs of the boat. Both Mike and Jimmy had a unique way of making building these massive boats look so easy, and for them it was because of their love of boats.[4]

Mike Tiliakos passed away in 1953 at the age of sixty-three. His devotion to his work and his family was inexhaustible. His grandson Nick remembers that even in the late stages of cancer, Mike would spend his days at the boat yard. He had a cot set up in a tool shed to rest on. Tiliakos Boat Building was sold in 1967, and Johnny Tiliakos left Fernandina to build boats in South America. Johnny completed eighty-four boats after his father's death (the first was christened the "Mr. Mike" in Mike's honor), before his own untimely passing at the age of forty-three.

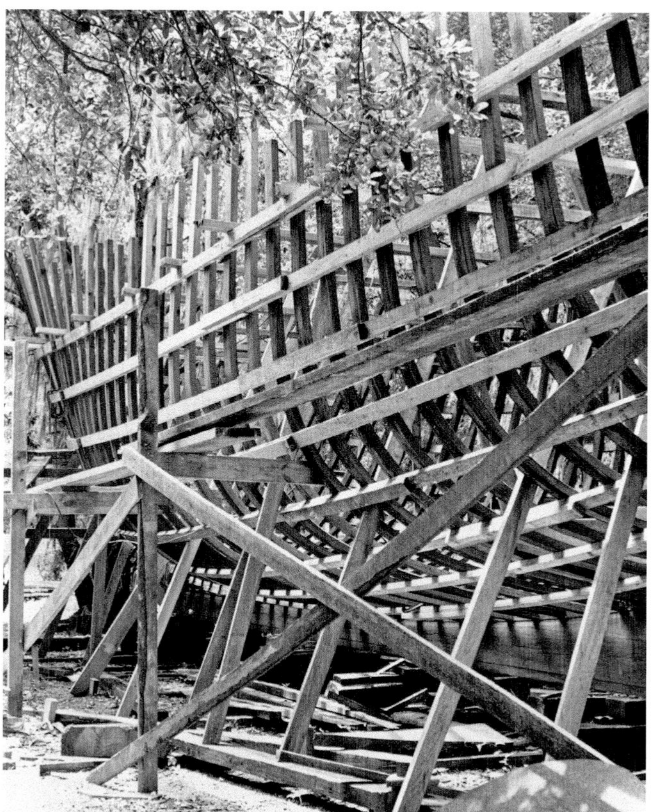

Constructing a hull at Deonas Boat Works. *AIMH Archives courtesy of Nick Deonas*

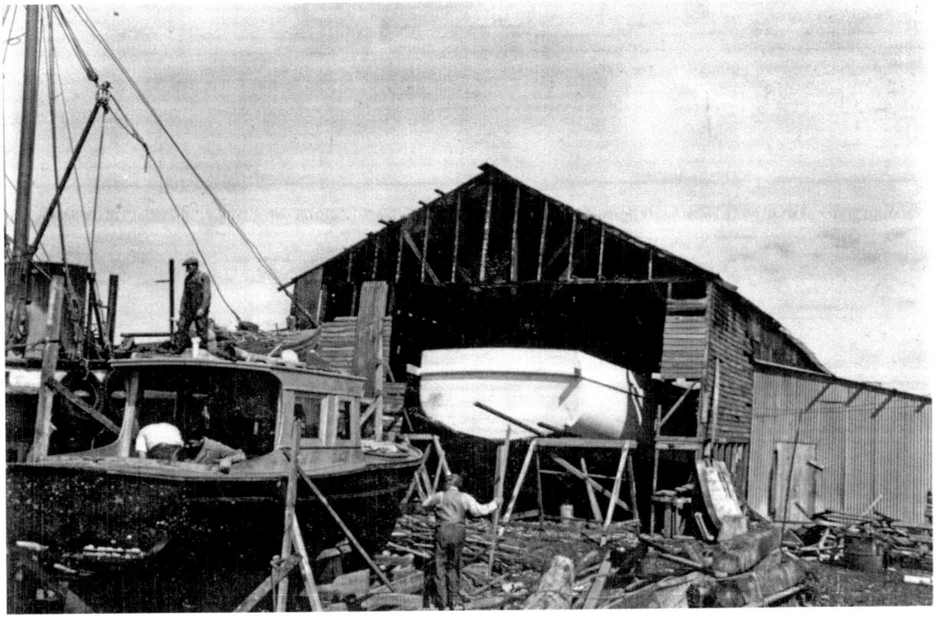

Both boatyards stayed busy. *AIMH Archives courtesy of Nick Deonas*

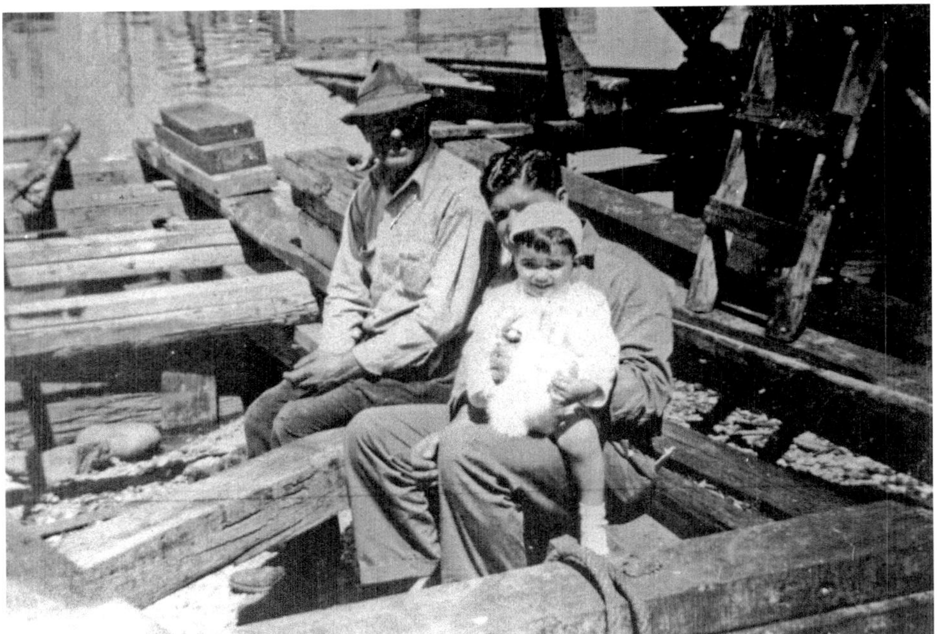

Mike Tiliakos sits with his son Nick Tilkiakos and grandson Nick Deonas. *AIMH Archives courtesy of Nick Deonas*

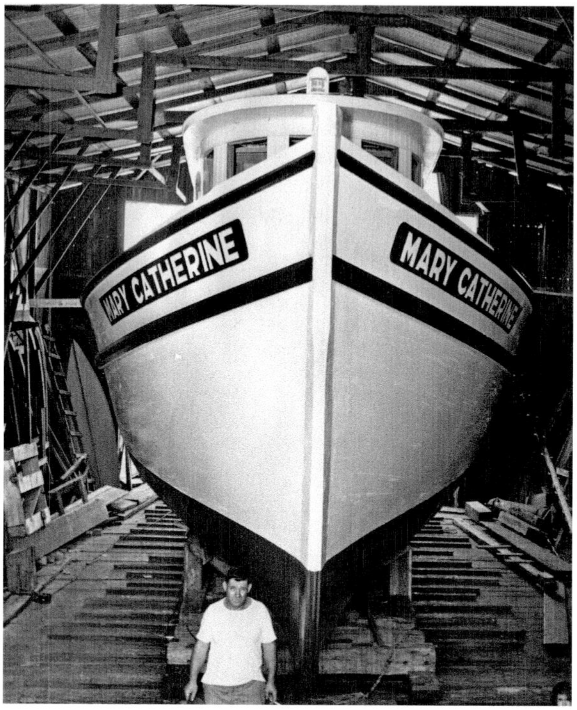

Jimmy Deonas with the *Nassau*. *AIMH Archives courtesy of Nick Deonas*

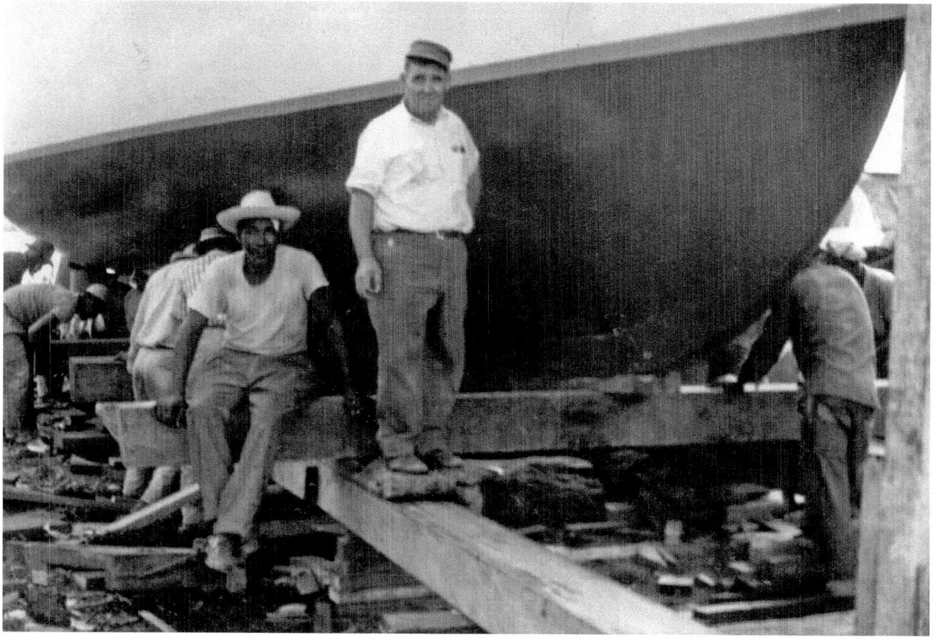

Deonas Boat and Motor Works in 1962. *AIMH Archives courtesy of Nick Deonas*

Unfortunately, a fire would destroy Deonas' boatyard on the island in the 1950s. But Jimmy was able to make a mutually beneficial deal with a friend who owned land in nearby Yulee, just off of State Road 17. The man knew that having Deonas working on his property would bring people to the area, so Jimmy moved Deonas Boat and Motor Works to the mainland and continued his work. Today there is a road in the area – Deonas Way – that bears his name.

Jimmy Deonas died in 1969, after devoting fifty-one years of his life to boat building. Jimmy passed his knowledge to his son Nick Deonas, who carried on his father's work until the boatbuilding portion of the Deonas' business closed in 1972. Nick became a fulltime fireman, retiring after 25 years with the department. He continues to preserve and perpetuate his family's boatbuilding legacy. The standard of craftsmanship set by Tiliakos and Deonas made Amelia Island a hub of quality boatbuilding for over five decades.

Boats in the Mediterranean style like the ones built by Tiliakos, Deonas, and fellow Greek boat builder Harry Xynides in nearby St Augustine, became what many people consider the archetypal shrimp trawler. These boats were incredibly seaworthy and were built to purpose. The design was widely copied and is still being imitated by boat builders all over the world.

6

Modern Shrimping: How it Works and Shrimp Facts

The basic procedure for most shrimp trawlers has remained relatively unchanged since the dawn of the modern industry. As a shrimp boat leaves the dock, the crew lowers the outriggers, which are the two long poles that give a shrimp boat its unique silhouette. The outriggers reach out over the water and keep the nets separated so they don't get tangled in each other. Next the nets are hoisted up over the side of the boat and lifted into position to be lowered into the water. Weights along the base of the net keep it dragging across the bottom, while a hood on top and floats keep the net open.

A smaller net called a "try" net is put in the water first to indicate if there are any shrimp to be caught in the area. If that yields any shrimp, the other trawls are lowered. The motion of the boat forces shrimp (and many other forms of marine wildlife, now limited by the introduction of exclusionary devices) into the nets. Pulling up the trawl too soon will mean missing out on shrimp, but leaving it in too long can damage the shrimp that are already caught. A good captain learns to tow the trawl for just the right amount of time.

The nets are then raised, and the catch is dropped onto the deck of the boat. The crew then go through and pick out healthy shrimp and any other usable fish. The rest – known as the "bycatch" – is swept overboard, and the shrimp are then rinsed off and stowed below deck on ice. Shrimp boats are followed by masses of seagulls and other birds (who often perch on the outriggers eying the catch and waiting for the bycatch to be discarded).

The time between catching and chilling the shrimp has to be kept as brief as possible to prevent spoiling. Some crews head their shrimp onboard as they go, while others will have their catch headed back at the dock. Heading shrimp (snapping off the head) helps with preserving the meat. This process continues until the hold is full or there are no more shrimp to catch in the area.

The boat returns to port and the shrimp are unloaded to be processed and sold. Ideally shrimp are processed at plants close to their port of landing. Shrimp are thawed by either air or water, and then peeled. In the modern era, machines are available to peel large, deep water shrimp, but most brown and pink shrimp are still peeled by hand. An experienced

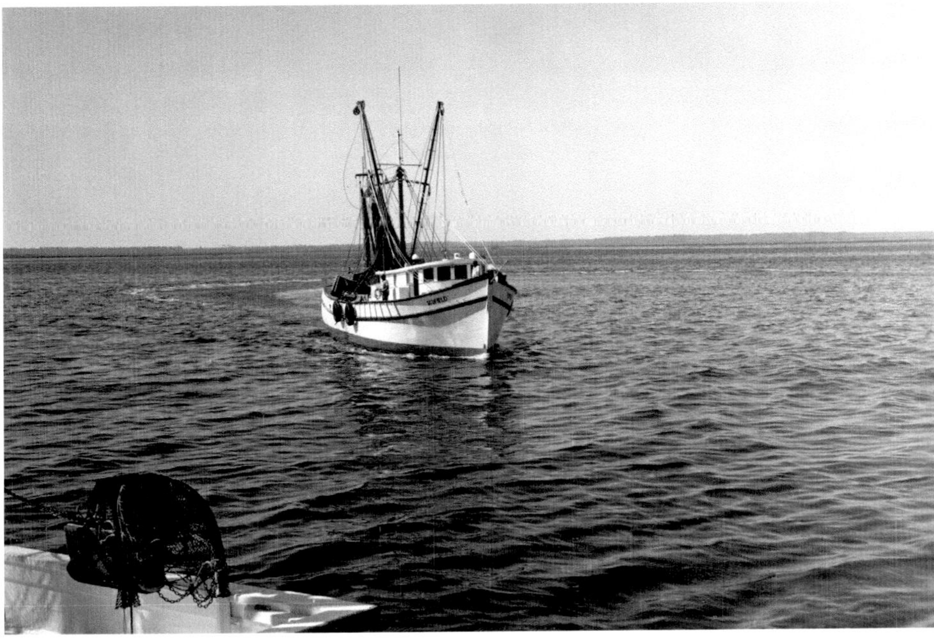

Two trawlers looking for shrimp near Fernandina—1965. *State Archives of Florida; photo by Karl E. Holland*

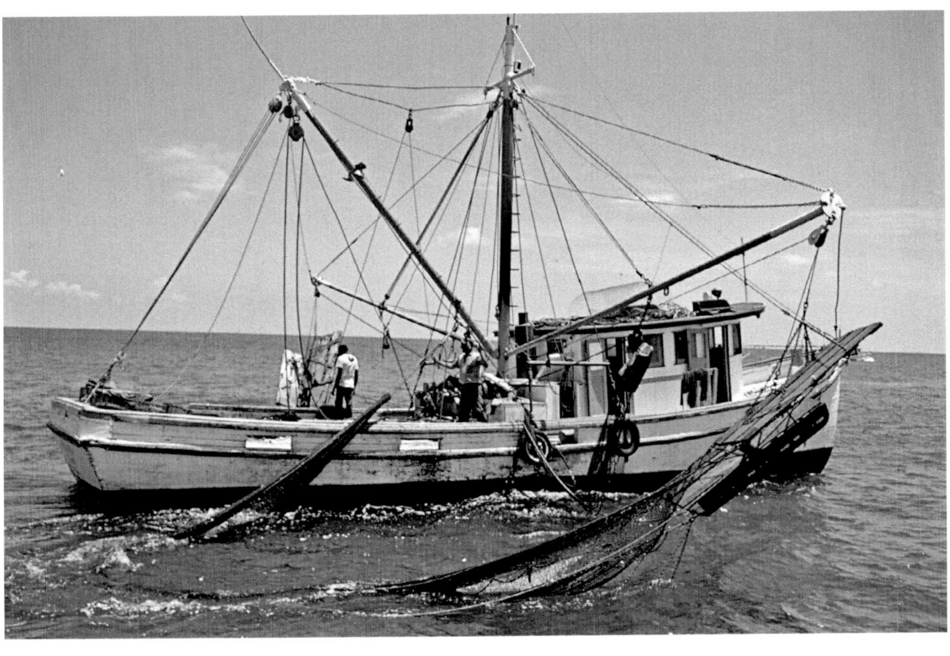

Hauling the trawls in. *National Oceanic and Atmospheric Administration*

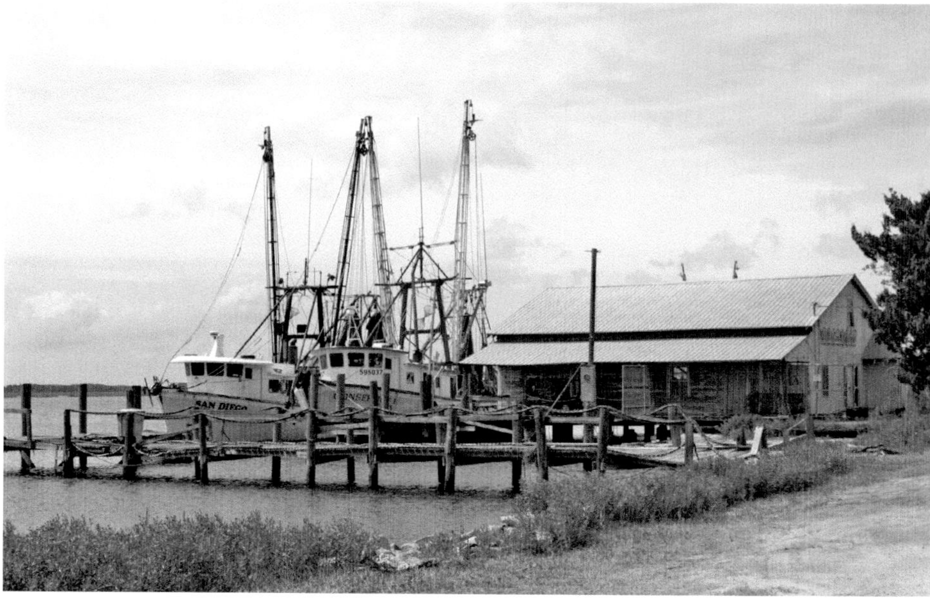

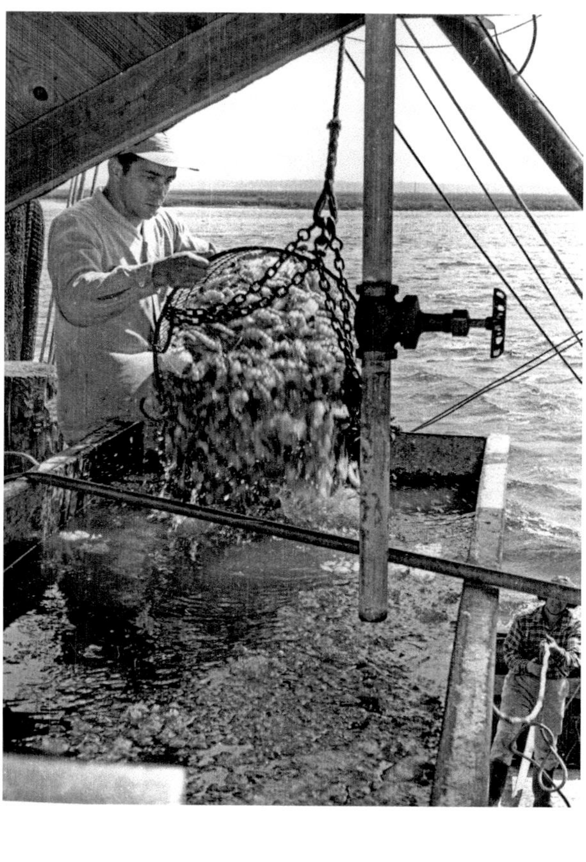

Above: Cook and Cook Seafood, Downtown Fernandina in 2015.

Left: Cleaning shrimp on the dock. *AIMH Archives*

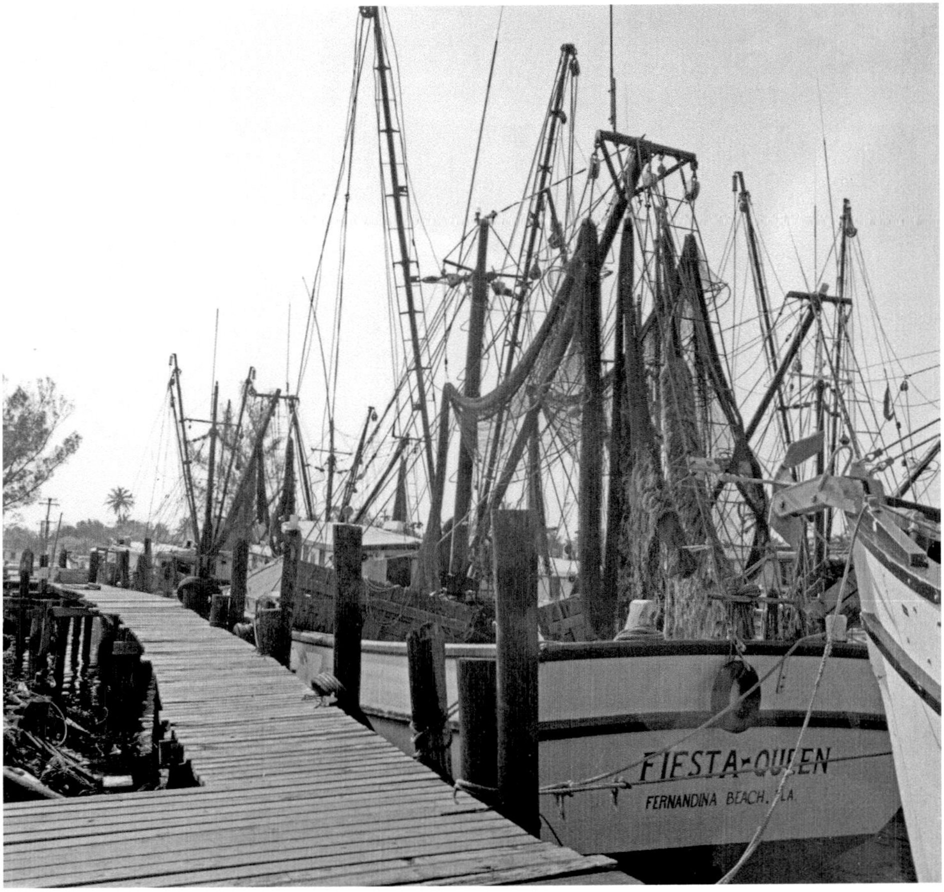

Shrimp boats docked at Fernandina, *c*. 1950s/1960s. *State of Florida Archives*

worker can head and peel around five or six pounds of shrimp an hour. The shrimp are then cleaned, and usually refrozen to be shipped and/or sold.

Most shrimp boat crews are fairly small, often only two or three people. The captain sets the course and steers the boat, but most also take part in the hard labor inherent to trawling. The workers who haul the lines, work the nets, and sort the shrimp are called "strikers." With overhead increasing at the expense of profits over the last few decades, many captains on small enough boats elect to trawl solo.

The word shrimp is used in some places interchangeably with the word prawn. There are more than 2,000 different species of shrimp in the world, around 300 of which are eaten by humans. Common colors include white, brown, gray, pink or yellow. Shrimp are crustaceans – relatives of lobsters, crabs, and crayfish – and they live in both salt water and fresh water. Some species of shrimp can grow up to 9 inches long, but most tend to be smaller. Female shrimp can lay up to a million eggs at once, which only take a few

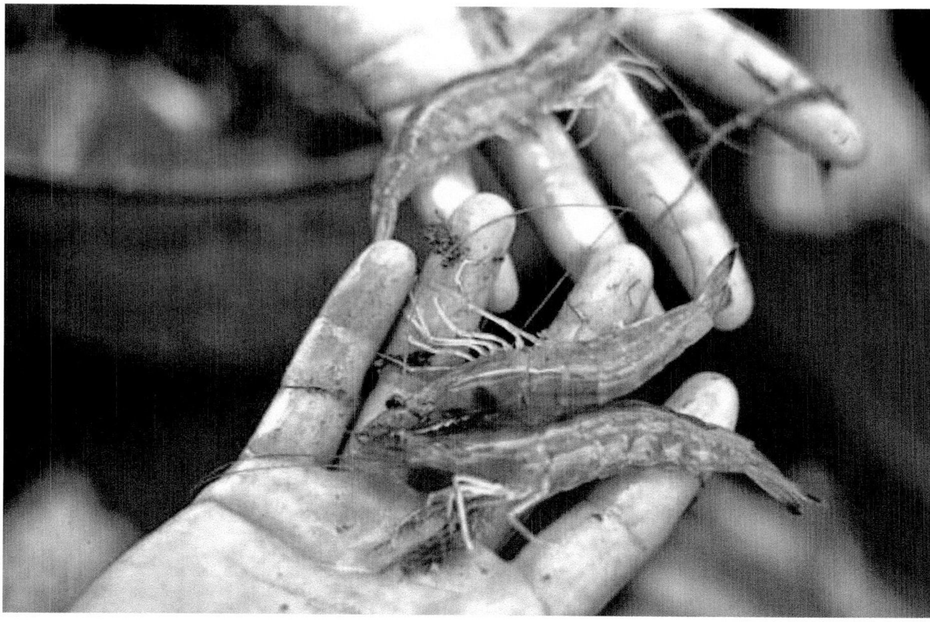

White shrimp caught near Fernandina. *State of Florida Archives*

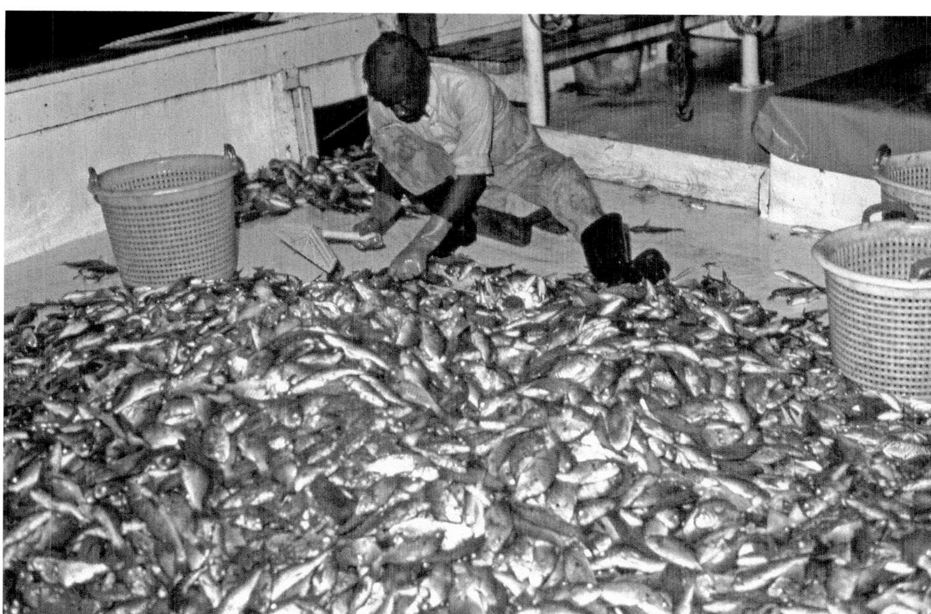

Local shrimper sorting the catch. *State Archives of Florida*

weeks to hatch. The life expectancy of shrimp varies depending on the species and habitat. The average lifespan is one to two years, though some live as long as six and a half years.

Shrimp are omnivores – they feed on a variety of plant and small animal species. Considered scavengers, they also consume plankton and detritus (matter produced by the decay or disintegration of an organic substance). Many shrimp are called "cleaning" shrimp because they remove ectoparasites from reef fish. They play an important role in the food chain, as a source of food for many other species including crustaceans, birds, whales, dolphins, sharks and humans. Shells from shrimp are used to make animal feed. Shrimp are a source of protein and selenium. They are also a source for vitamins B12 and D, iron, and zinc.

There are three species of shrimp that are the most commercially important: pink, brown, and white. All three species can be found on either coast, but prefer different habitats. They all use estuaries as nursing areas, and frequent seagrass beds and algal mats during juvenile stages. White and brown shrimp spawn are usually found in shallow water with a muddy bottom. White shrimp favor water with a low salt content, while brown shrimp prefer high levels of salinity. Pink shrimp are found in a wide variety of salinities.

About eighty-four percent of the shrimp eaten by Americans is imported. A substantial portion of that comes from the aquaculture farms in Asia and South America. Americans spend over $4 billion on imported shrimp annually, making shrimp the most valuable imported seafood product in the US. Americans consumed a total of 1.4 billion pounds of shrimp in 2012, and that figure is trending up.

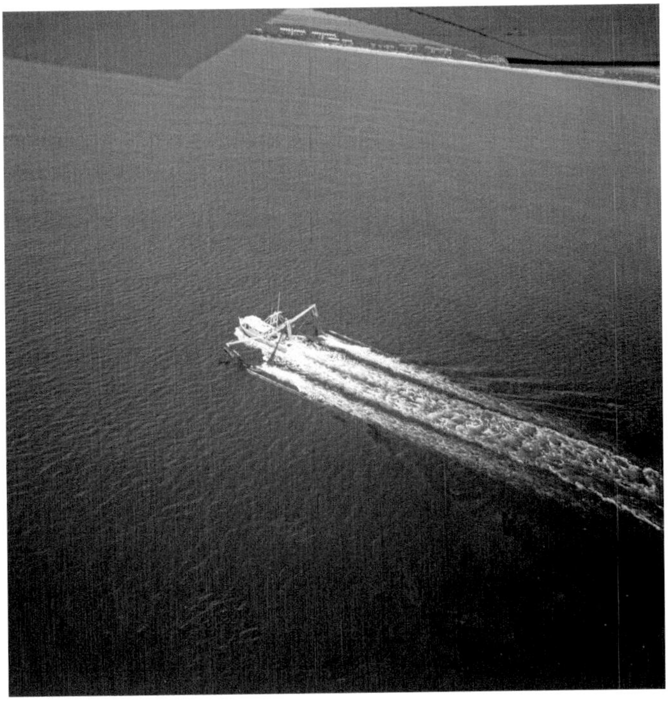

Aerial view of a shrimp boat in action—1989. *State Archives of Florida; photo credit: Overton*

7

Net Making

HISTORICALLY SHRIMP HAVE BEEN caught with cast nets, or haul seines (also, a simple fishing net that hangs vertically in the water, with the bottom of the net held down by weights and the top kept on the surface with floats). The seine could be cast from the shore or from a boat. The ends of the net would be drawn together and pulled towards shore. Before the introduction of power boats, pulling in a seine could be extremely hard work, depending on the size of the net.

In the earliest years of the shrimping boom, most shrimpers sewed and repaired their own nets. A few people in the area specialized in making shrimp nets, like James W. Delaney. The Delaney family has a long and esteemed history in Fernandina. Thomas and Sarah Elizabeth Delaney (sometimes spelled Delany) were literate former slaves who made their way to Florida during the Civil War. One of their sons, Ray Samuel Delaney, later became Fernandina's City Commissioner, and another son, Henry Beard Delaney, became a bishop – establishing a tradition of political and religious service in the family. For many years, Mr. Delaney and his family sewed high quality nets from the front porch of his home on Beech Street. The Delaney's made shrimp nets that were protected from sharks tearing into the nets. His descendants continued crafting nets until the early 1980s.

A Fernandina resident named Bill Burbank would take the net making baton and run with it, starting a local business that would eventually supply nets to shrimpers all over the world and become the industry standard. William Hunter Burbank was born on Cumberland Island in 1893. William (known to locals as "Cap'n Bill") was a fisherman, who moved back to Fernandina during the bluefish run around 1915. After the bluefish played out, he became active in the still nascent shrimping industry. Cap'n Bill married Effie Peterson, who came from a family of skilled fishermen originally from Southport, North Carolina (her brothers included Bluff, Sloan, and H. R. "Hydie" Peterson). The couple had six children: William Jr, Edith, Lucille, Dorothy, Frank, and Sharlene.

Like most boat owners at that time, Cap'n Bill made his own nets. Growing up on Cumberland he'd seen the nets used by bluefish fishermen from North Carolina. He noticed that these nets had especially small holes, which often caught shrimp incidentally. With that

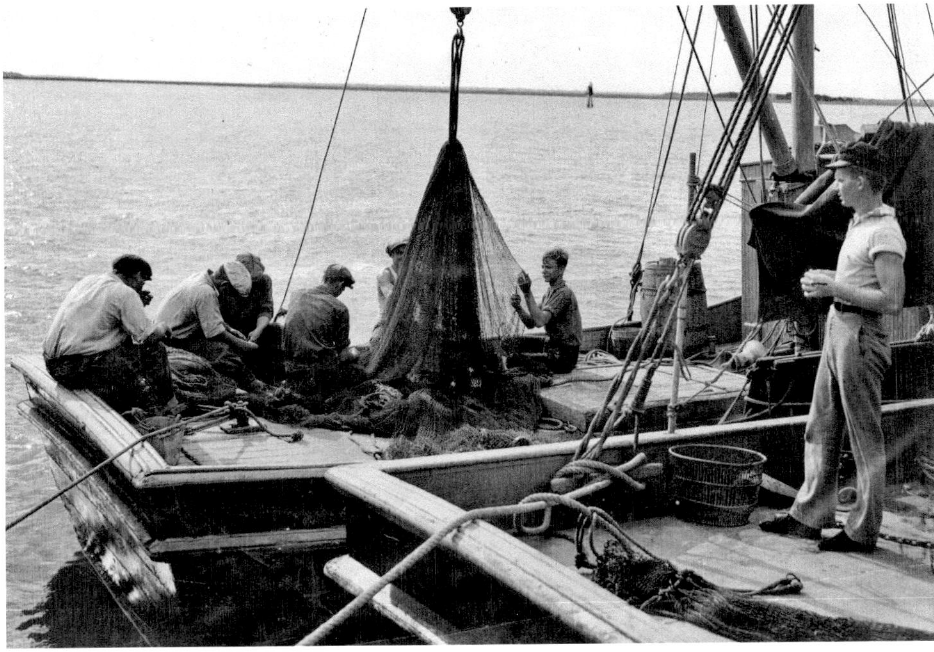

Shrimpers mending nets at Fernandina. *AIMH Archives*

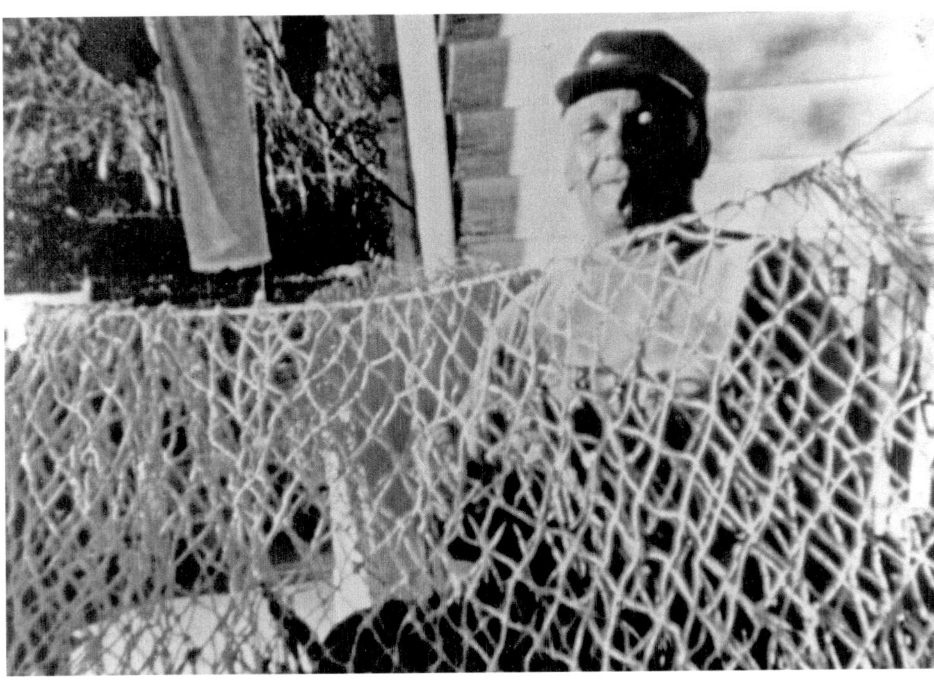

James W. Delaney Jr. (1903-1977). *AIMH Archives*

in mind, Bill became adept at making trawls (a bag-like net that is drawn behind a boat) and began refining his own designs – adding a wing to make the net taller and increase the catch. As other fisherman noted the success of Cap'n Bill's trawls at catching shrimp, they asked him to make nets for them as well. Bill sold his boat to his brother in law Bluff Peterson and began devoting most of his attention to net making – constructing them underneath his home on South Fletcher Avenue. With a handshake agreement, Burbank Trawl Makers became exclusive subcontractors to Standard Marine Supply Corporation.

Burbank Net Makers was a family business from the start, and Cap'n Bill passed his skills on to his children and grandchildren. After serving in the Navy during WWII, William Hunter Burbank II joined his father and his brother Frank in the net making business. Billy II (affectionately known as "Cap'n Daddy") eventually became owner and president of Burbank Trawl Makers Inc., and in 1957 the business moved to a waterfront location on N. 2nd Street. Cap'n Bill continued to sew nets for the company until he passed away in 1975. Billy II grew the business by improving upon existing designs, and implementing a few of his own. His sons learned net making at an early age and Billy III, Tommy, David, and Johnny eventually all joined the business.

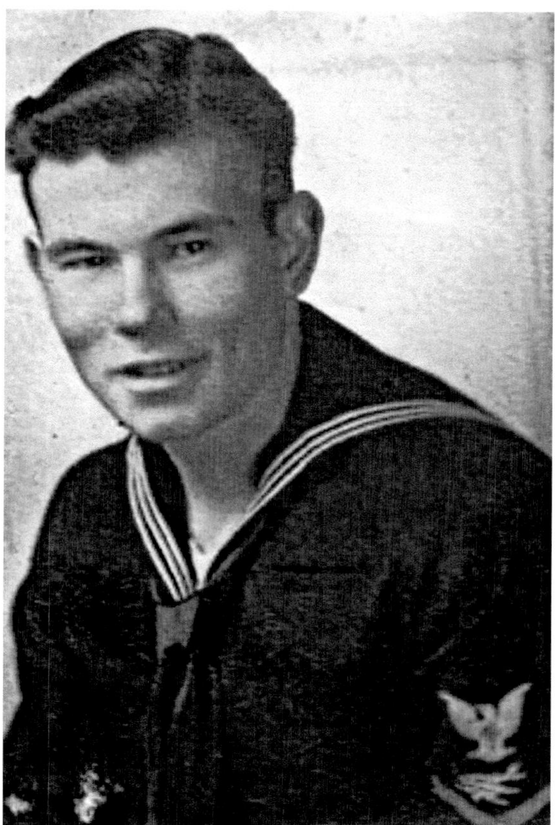

William Hunter Burbank II. During World War II, Burbank served on the U.S.S. *Tennessee.*

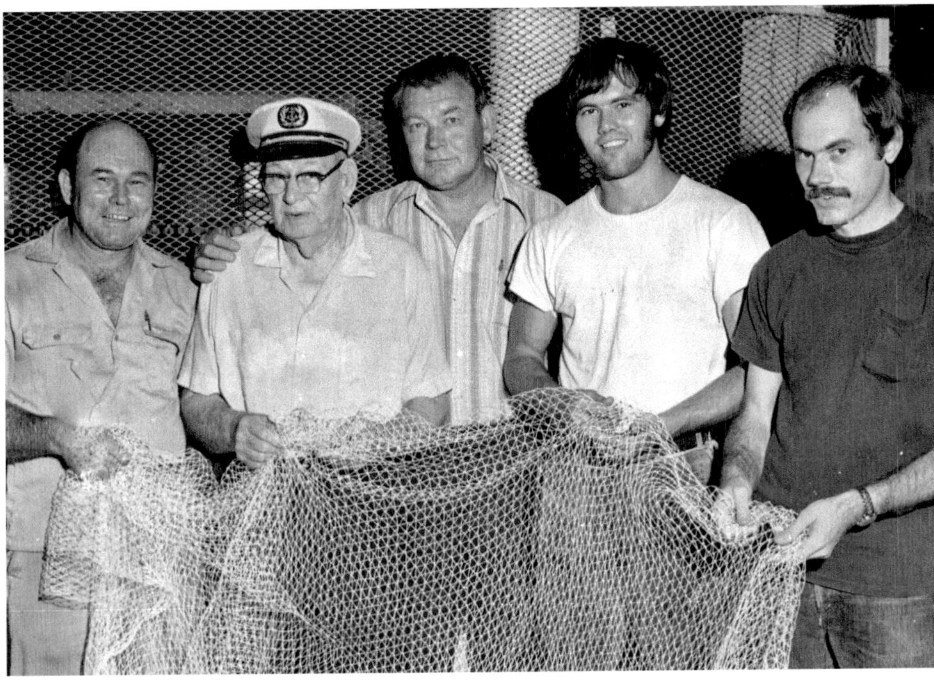

From left to right: Billy Burbank II ("Cap'n Daddy"), Billy Burbank I ("Cap'n Bill"), Frank Burbank, Billy Burbank III, and Charles Jones in 1971. *AIMH Archives*

Frank Burbank moved to Tampa in the late 1950s and continued making nets for the Hardees. He also owned shrimp boats, following the migration of shrimp to Texas each winter. Frank's son Frank Burbank Jr. grew up sewing nets and shrimping. He attended Texas A&M, and after several years as a port captain in the pogy fishing industry he returned to Fernandina to work for Standard Marine Supply (continuing the Burbank/Hardee connection).

By 1971 Billy Burbank III had joined Burbank Trawl Makers full time, spending over a hundred hours a week in the net shop. A few years later he bought the company from Billy II, who stayed on sewing nets for several years, just as William Hunter Burbank I had done before him. Building on Cap'n Bill's legacy, the Burbank family continued to make innovations in net design. They developed turtle exclusion devices (known as TEDS), which use an aluminum grid set at a forty-five degree angle, allowing sea turtles and other large animals to pass through the bottom of the net. The Burbanks have also cultivated designs featuring fish portals, which allow smaller fish to escape the net.

By the late 1970s government intervention and rising fuel prices had taken their toll on the shrimping industry, and local businesses who depended on shrimping began to feel the effects. Billy Burbank III saw the writing on the wall and began looking for new ways to evolve and expand the business. Opportunity came in the form of sports nets. The Burbanks built their first batting cage in 1979 for Orange Park High School's baseball coach Jack Rhine.

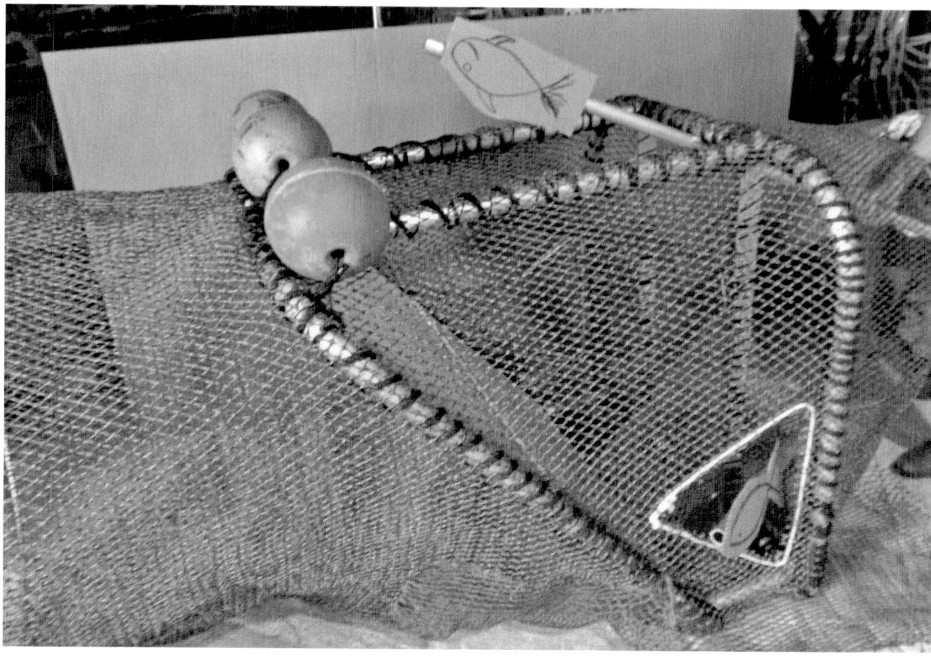

Demonstration of a Turtle Exclusion Device/Fish Portal in action at Burbank Sports Nets. *Photo courtesy of Billy Burbank*

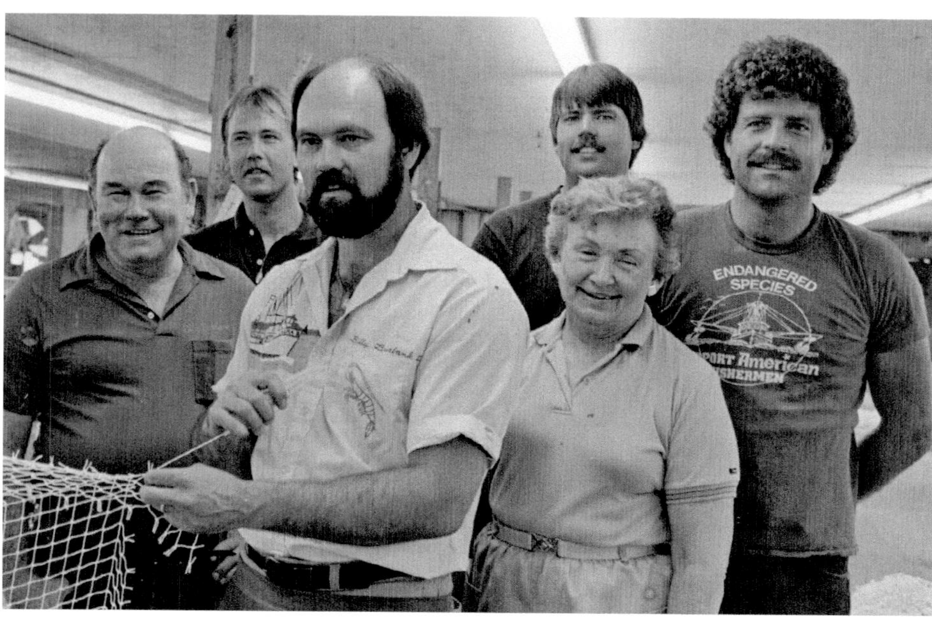

Left to right Billy Burbank II, Tommy Burbank (back row), Billy Burbank III, Johnny Burbank (back row), Lucille Barbee (sister to Billy, Jr.), *David Burbank Photo Credit: AIMH Archives*

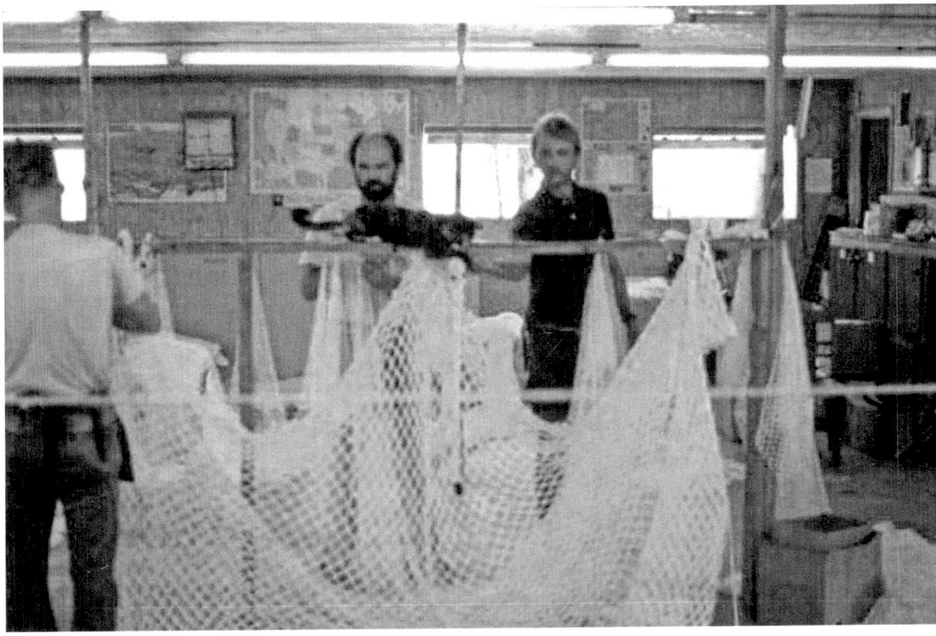

From left to right: Billy Higginbotham, Billy Burbank III, and Tommy Burbank work on a net in the Standard Marine building. *Photo courtesy of Billy Burbank*

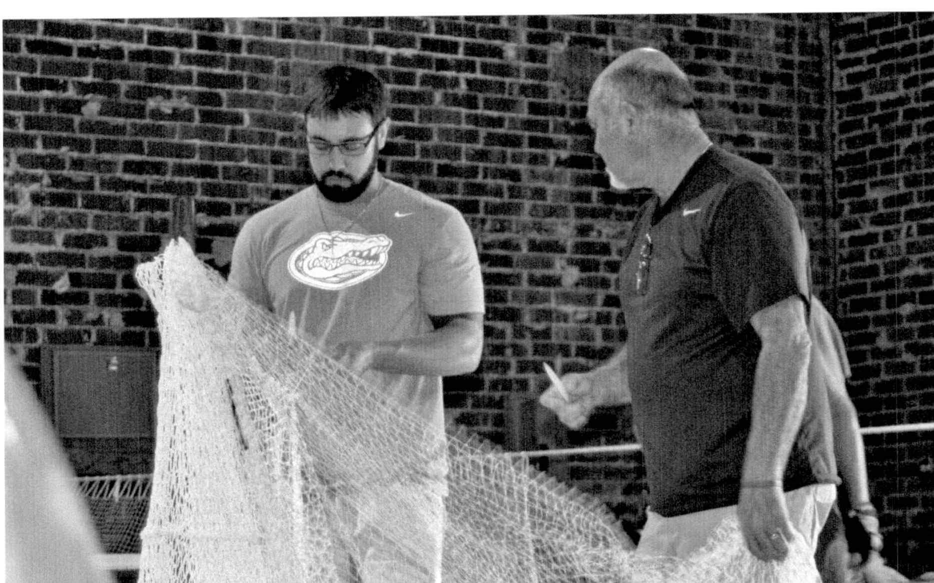

Billy and Hunter Burbank sewing nets in 2015. *Photo by Author*

Rhine later became head coach at the University of Florida and asked the Burbanks to install their nets at the Gator's ballpark. The Burbanks quickly expanded past baseball into making nets for other sports and for collegiate and professional athletic programs across the country.

In the early 1980s they revolutionized the sports net industry by introducing UHMWPE fiber (a strong and lightweight fiber the company originally developed to help shrimp boats conserve fuel in the 1970s) to the stadium of the Chicago White Sox. Other organizations followed suite, and Burbank nets can now be found in over 75 percent of Major League facilities: including the Atlanta Braves, the St. Louis Cardinals, and the San Francisco Giants. As the company's president and owner, Billy Burbank III rebranded the company as Burbank Sports Nets in 1993.

In 2004 The Burbanks moved to the former Pogy Plant (owned by the Corbett family) the former center of Fernandina's Menhaden fishing industry. The shop continues to produce thirty to fifty nets per week, not only for sports facilities, but for shrimpers around the world. With William Hunter Burbank IV (who goes by "Hunter") in charge of the sports net side of the business, the company has now been in the family for four generations, and endured for nearly a century. As the Burbanks aptly point out, "the sun never sets on a Burbank net."

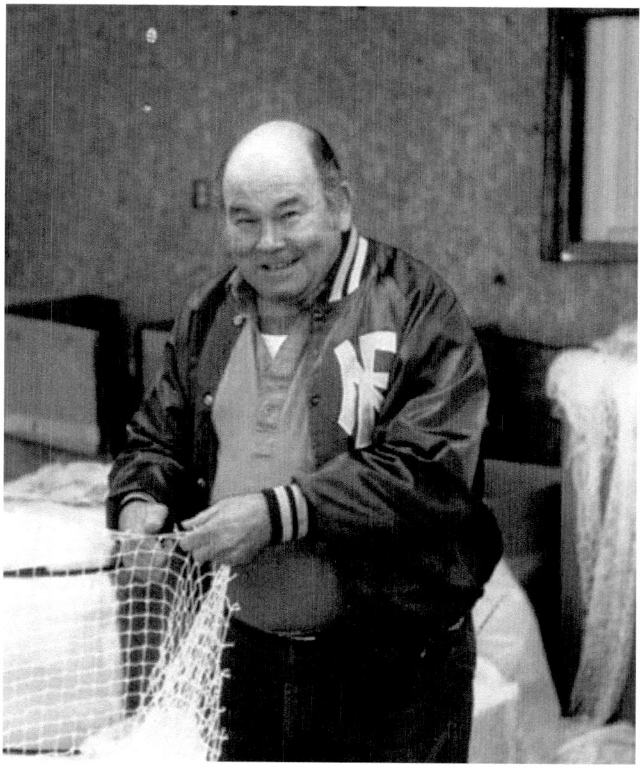

Billy Burbank II working a net. *Photo courtesy of Billy Burbank III*

8

A Head for Business

Several local entrepreneurs who had already made a name for themselves in local business couldn't help but see opportunity in the nascent shrimping industry. For anyone with the capital, it was a natural transition to invest in some aspect of shrimping, from owning a boat, to establishing a cannery. As the forefront of the shrimping industry moved further south and to the gulf, many Fernandina businessmen remained at the head of the charge.

From the 1910s to the 1940s, when local production remained high, there could be around a dozen fish houses for processing shrimp in operation on the island at a given time. Several prominent local families saw an opportunity and branched out into the processing side of the business. The Salvador, Versaggi, Hardee, Vichio, and Chadwick fish houses lined the waterfront, later joined by names like Williams, Smith, Merror, Davenport, and Peterson.

Prominent local entrepreneur Albert Hirth founded the Atlantic Fish and Shrimp Company in 1918 as a market for local shrimp. Shortly after WWI, Hirth sold the business to his brother-in-law B. B. Brooks, who renamed the company Brooks Seafood Company and established shrimping canneries in Fernandina, St. Augustine, Beaufort, South Carolina, and Galveston, Texas. B. B. sold the Fernandina operation to his brother W. M. Brooks in the mid-twenties and went on to become one of the early leaders of the Louisiana shrimp rush. [5]

In 1936, W. M. Brooks spent $12,000 updating his factory, making it one of the most state of the art facilities in the South at that time. A detailed account of his operation was given in the October, 1936 edition of *The Fish and Oyster Reporter,* which provides a time capsule of the canning business at that time. "The plant, which covers an area of 40 to 140 feet, when operating at full capacity will have about 140 workers on a payroll ranging daily from $200 to $300. A regular fleet of 15 to 20 shrimp boats will enable the Brooks Company to turn out 450 cans and 50 cases of jars hourly."[6]

Mr. Brooks and his foremen were vigilant in maintaining cleanliness and quality control to protect their international reputation for quality shrimp. As soon as they were brought into the factory, the shrimp were loaded into a ten foot long washer made of galvanized iron. The shrimp were then inspected on a conveyer belt, while in route to a second washing.

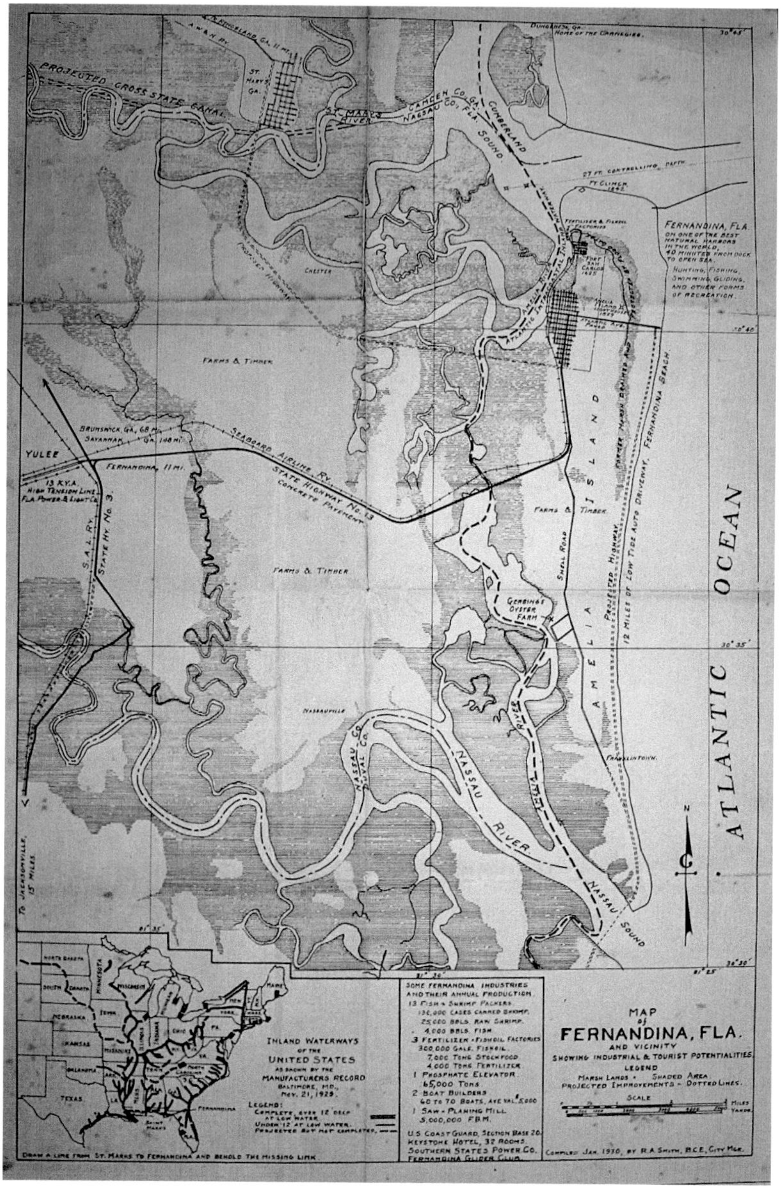

1930 Map of Fernandina, Fl and vicinity showing industrial and tourist potentialities along with some local industries and their annual production including: 13 fish and shrimp packers, (130,000 cases canned shrimp, 25,000 bbls. Raw shrimp, 4,000 bbls. Fish), 3 fertilizer and fish oil factories (300,000 gals. Fish oil, 7,000 tons stockfood, 4,000 tons fertilizer, 2 boat builders (60 to 70 boats, average value $5,000.00). Compiled by R. A. Smith, B.C.E., City Manager. *AIMH Archives*

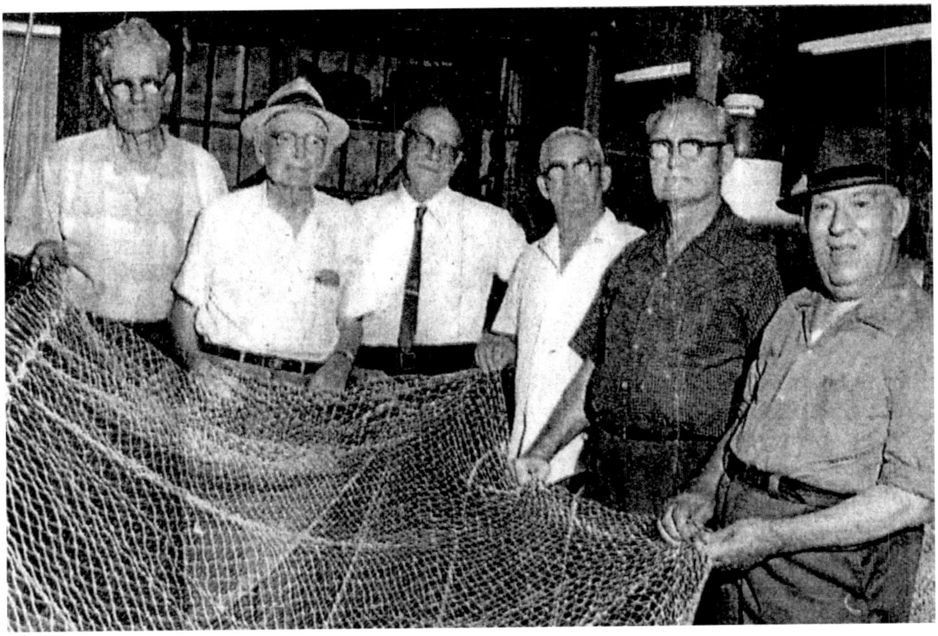

Some of the elder statesmen of local shrimping, from left to right Emmett Freeman, an unidentified man, D. A. Dave Cook, Capt. Dave Tapper, Sloan Peterson, and Charlie Susan.

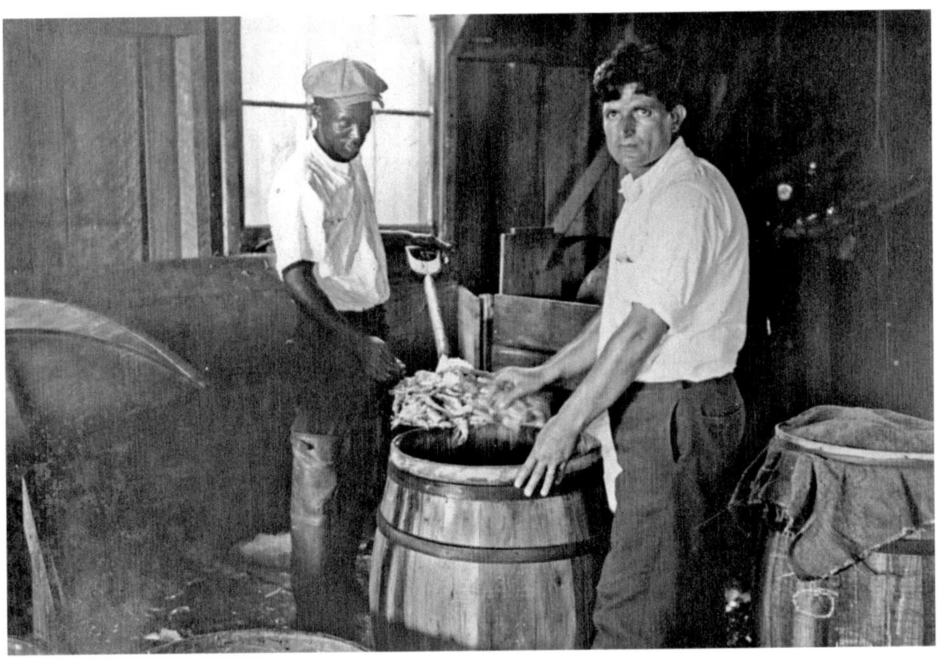

Packing Shrimp at Fernandina. *AIMH Archives courtesy of Billy Burbank III*

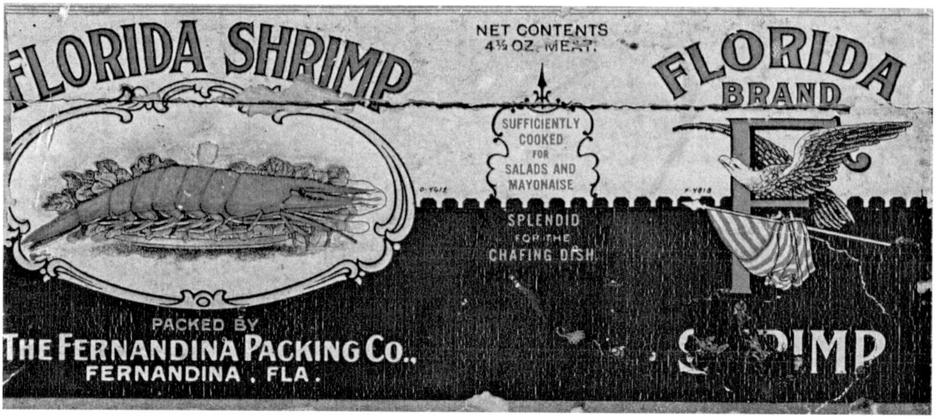

Label from the Fernandina Packing Company, managed by Frank Gorenflo. Gorenflo's company records from May of 1922 through December of 1924 currently reside in the archives of the Amelia Island Museum of History, and are full of transactions with recognizable names in the shrimping industry.

After that, they were put on ice and stored overnight. The next morning, the shrimp were poured onto three iron tables, around which, up to a hundred and ten workers would stand peeling and heading the shrimp. The tables were equipped with running water around both sides, which washed the peelings and heads into the Amelia River.

Next the shrimp were carried on a flume of water for a third washing and another inspection. They were then packed into fifteen gallon cans to be brined and boiled. The shrimp were sorted by size and inspected a final time before being packed into cans or jars by machine. The jar capping machine could seal forty jars a minute, while the canning machine could cap forty-eight cans in the same amount of time. The Brooks cannery carried on into the early 1940s, still packing shrimp into their iconic glass jars.

Brooks' establishment at Front Street and Dade later became the site of Hydie Peterson's shrimp dock. Hydie R. Peterson Jr. was the nephew of Effie Burbank (wife of net maker William Hunter Burbank, and a part of the Peterson family that included so many capable fishermen. Fresh out of the US Navy, Hydie took a job with Standard Hardware Company, before branching out into the retail seafood business. Peterson was the proprietor of the Peterson Packing Company, Seafood City, Sportsman's Wharf, and Cap'n Pete's. For many years his seafood enterprises were an institution in Fernandina, and one of the best outlets for buying wild caught shrimp.

One of the most complete success stories in local shrimping belonged to Scotsman John C. Fergusson. After immigrating to the United States, Fergusson made his way to Fernandina and began working on shrimp boats in 1922. Starting from the bottom, he worked his way up from striker to captain, and by 1939 was the owner of three boats of his own. From there Fergusson doggedly worked to build one of the largest fleets in the shrimping industry.

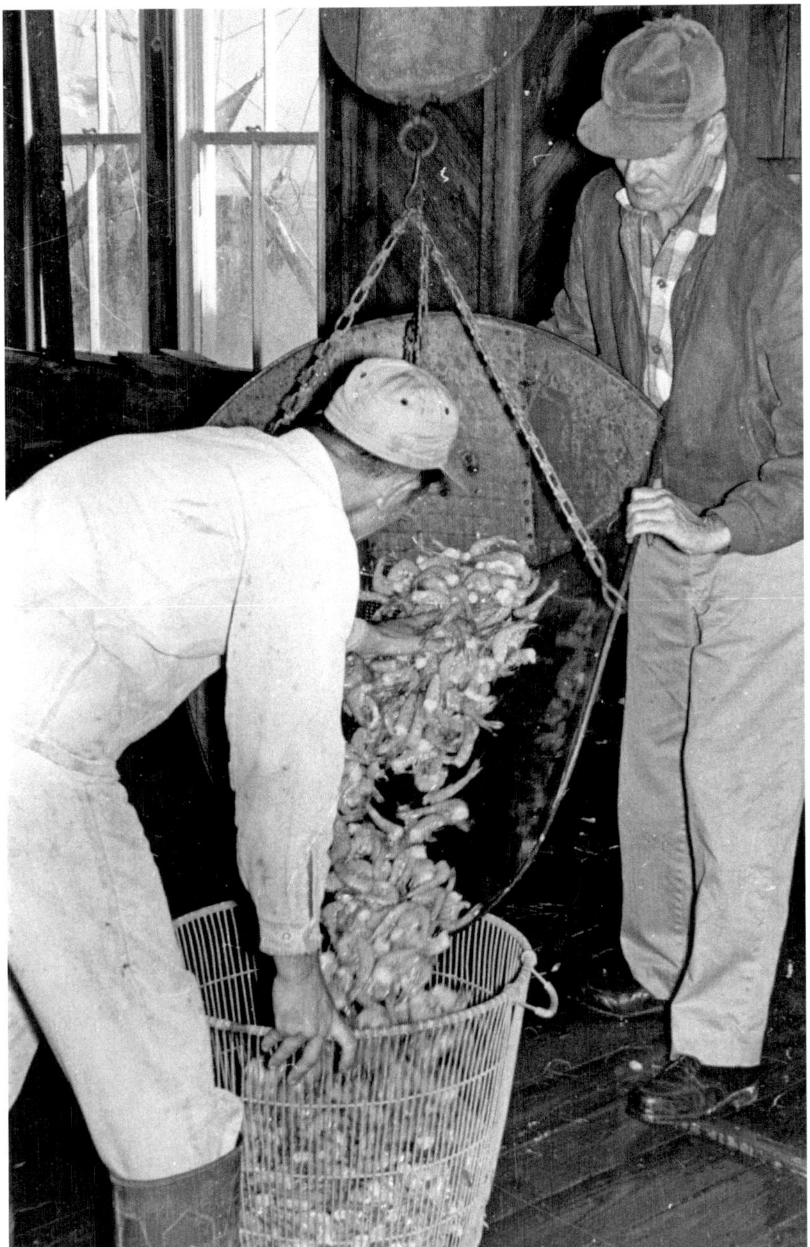

Processing shrimp in Fernandina. *AIMH Archives*

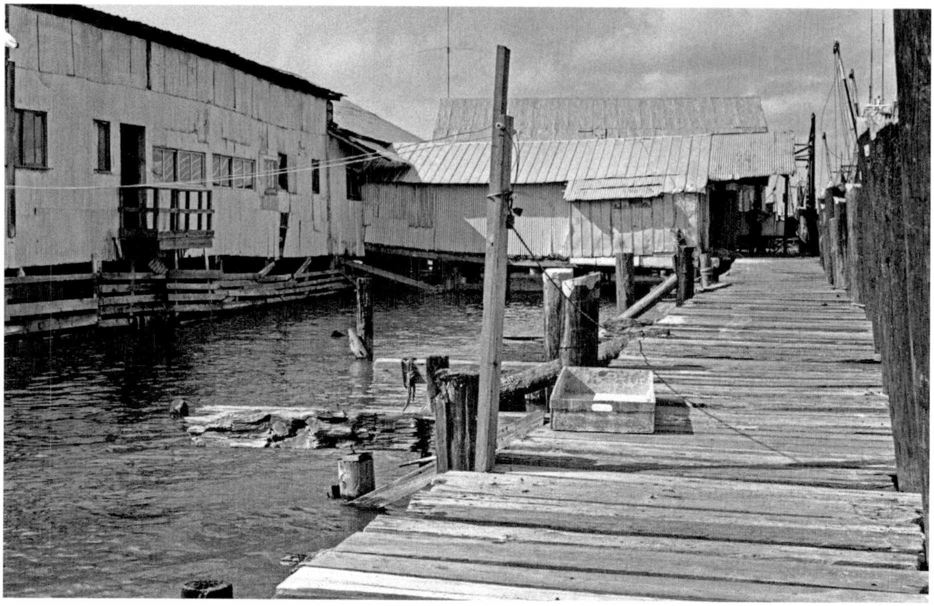

Hydie Peterson's shrimp dock (former site of W. M. Brooks' Cannery) at Front Street and Dade Street. *AIMH Archives, courtesy of Mrs. H. R. Peterson*

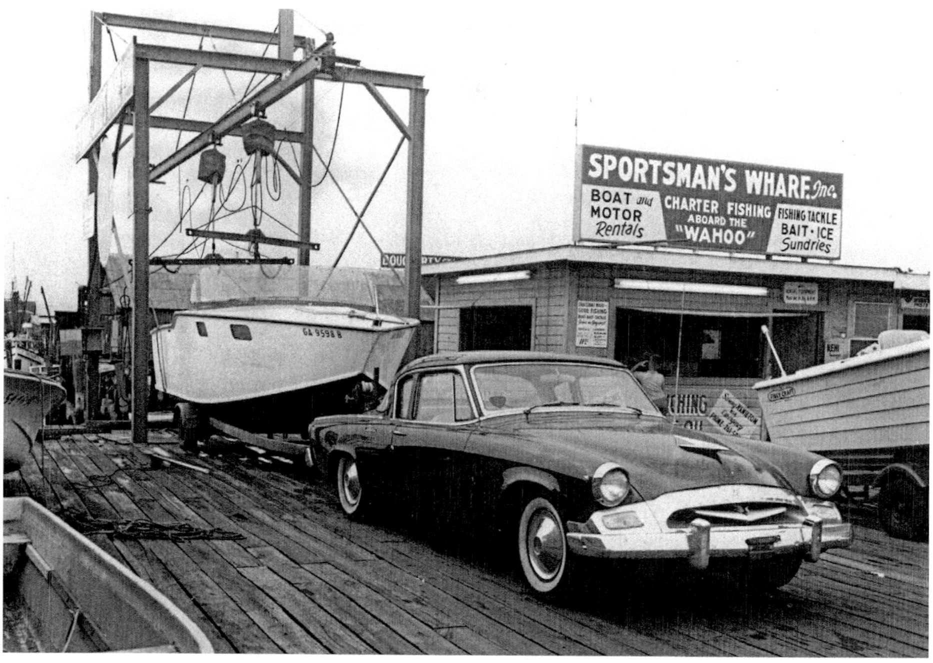

Sportsman's Wharf. A seafood market and a bait and tackle shop owned by Hydie Peterson. *AIMH Archives*

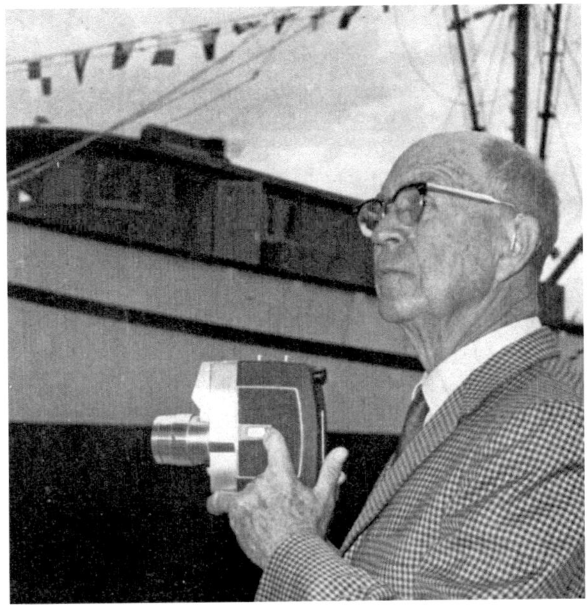

Harry F. Sahlman. *AIMH Archives; photo by David B. Lord*

Businessman Harry F. Sahlman was originally in the lumber business, the owner of Fernandina Lumber Supply Company. Sahlman made a deal to supply lumber to a local boat builder for the construction of a shrimp boat. Facing the hard times of the depression, the boat didn't sell and Sahlman found himself the owner of his first shrimp boat, which he dubbed the "Jack and Jill." Sahlman started a seafood company in Fernandina in 1936, and within a few years he had a fleet of fourteen boats. In his first year in the business, Sahlman's boat the "Mickey Mouse" caught a record seventy-five bushels of shrimp in one trip (the next closest contender was the "Fortune" owned by S. Versaggi and Sons, which brought back fifty-five bushels in one outing).

Sahlman's son Charles "Jack" joined the company, which followed the shifting industry to Tampa, and then into South America. Sahlman Seafood owned a dock, processing plant, and repair yard in British Guyana, along with many other holdings. Boats in the Sahlman fleet were among the first, and the most successful firms to fish the lucrative Key West run of 1955. The Sahlmans helped pioneer shrimping in the waters of Mexico, and both South and Central America using the methods born in Fernandina.

Sahlman Seafood sold their shrimp on the East Coast under the famous "Bee Gee" brand. By 1958 the fleet had grown to 150 boats. In the 1970s the Sahlman's innovated the use of refrigerated "freezer" boats, which replaced the traditional practice of simply storing shrimp on ice. In 1997, the Sahlmans made the transition to aquaculture, opening their own shrimp farm in Nicaragua. The facility now features forty-four ponds, which can produce up to 120,000 pounds of shrimp a day. Sahlman Seafoods celebrated its 75th anniversary in 2011, and that same year won the Secretary of State's Award for Corporate Excellence.

In 1964, Salvatore Tringali's sons Joe, Angelo, Sebastian, and Tony built a fish house they called Tringali Brothers Shrimp Co. Until that point the family had rented space in the Tiliakos boat yard to dock their boats. The Tringali fleet continued to operate out of Fernandina until the 1980s, when they sold their four remaining boats. Unfortunately, a fire in July 1985 destroyed the Tringali Brothers Shrimp Co. building.

D. A. "Dave" Cook was the head of a prominent local shrimping family, which produced generations of proficient shrimpers. D. A.'s four sons Clarence Hinton, David "Junior," Ed , and John all grew up to be shrimpers. The family built up their own fleet of shrimp boats, maintaining docks in Nassauville and Fernandina. By the early 1980s, D. A.'s sons owned ten trawlers, and several of his grandsons also became shrimpers. Though Fernandina's shrimping industry has slowed down considerably, the Cook family still owns shrimp boats and docks in the area.

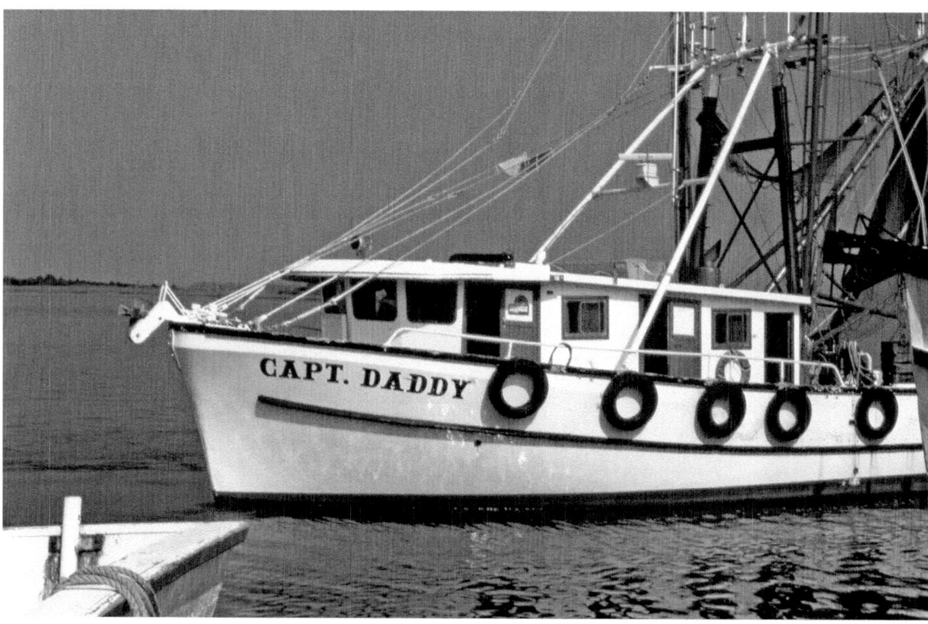

The *Capt. Daddy* in 1980, captained by David Cook (grandson of D. A. Cook. *State of Florida Archives*

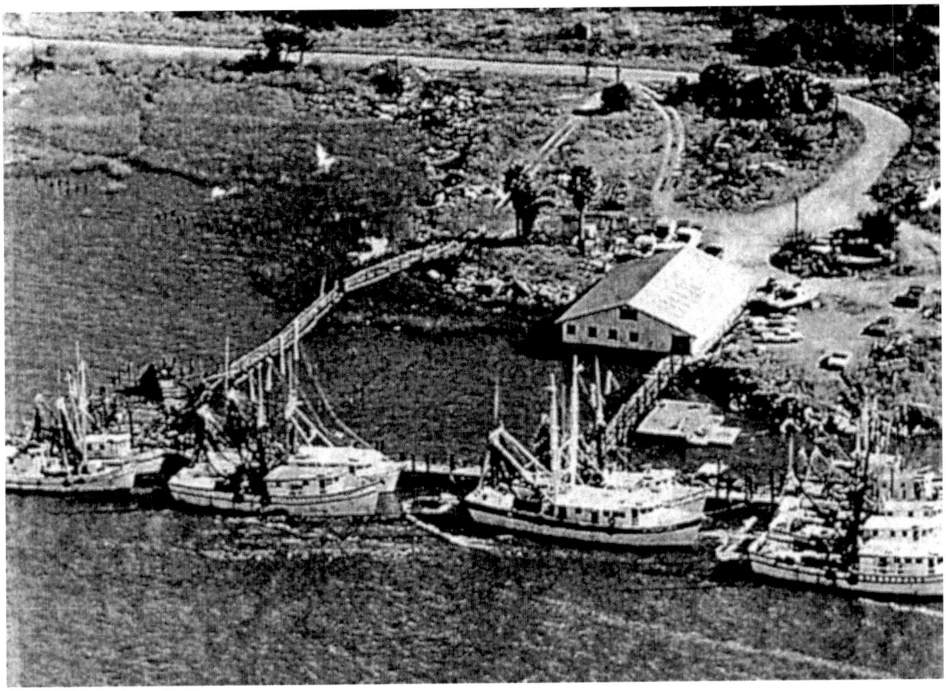

The Cook's Shrimp Boat dock downtown, *c.* 1980.

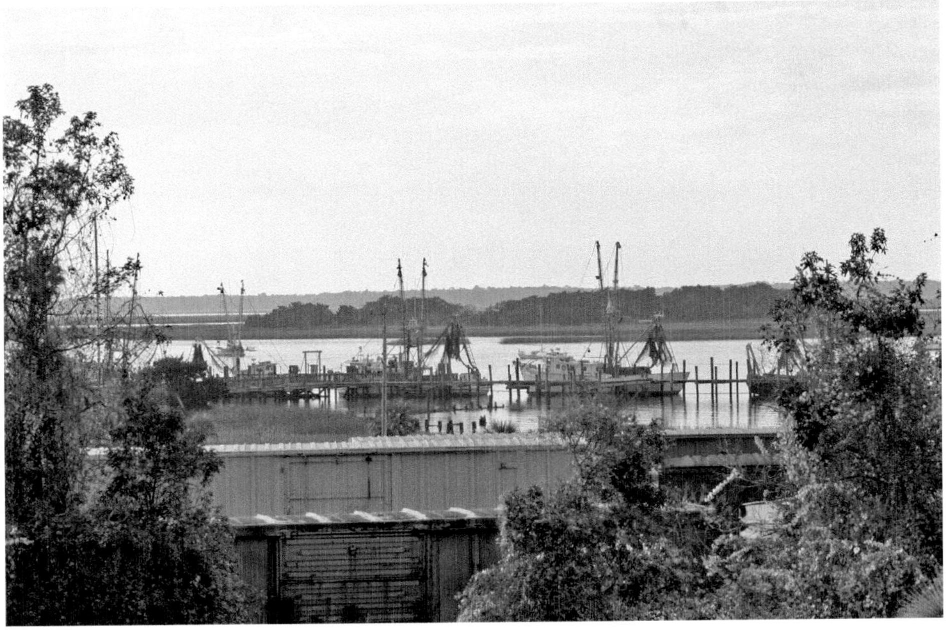

The Cook's dock in 2015. *Photo taken by the Author*

9

Carving a Life Out on the Water

It is important to understand that African American fisherman and workers were crucial to every aspect of the seafood industry in Fernandina and the surrounding areas. In the days of segregation and unjust practices of hiring and compensation, shrimping – along with other types of commercial fishing – was one arena where a person could earn a living based on merit. As opposed to many other ways of making a living at the time, commercial fishing provided a person with the opportunity to make good money based on skill and production, relatively uninhibited by the color of their skin.

By the time shrimping began to boom in Fernandina, Old Town – the historic center of Fernandina – had become the heart of the African American fishing community (there was also a significant number of black families living in the Five Points area on the west side of the island). People who lived in Old Town at that time talk about the tremendous sense of community residents had back then. Joseph Perkins (forty year veteran of local shrimping) recalled:

> Everybody got along. Everyone looked out for one another. People that didn't have, we'd give to. Everybody shared back in those days.[7]

Annette Myers (daughter of shrimper Capt. Mac McCollough and wife of shrimper Elmo V. Myers) said:

> Seafood was plentiful. No one went hungry. You could always go to the docks and get enough handouts or scraps of seafood to feed your entire family. There were a lot of boats around and a good living to be had, even as a crew member.[8]

There are several families of African American descent – like the Traeyes, the Richos, and the Bennetts, along with several others – whose involvement in local shrimping goes back to the earliest days of the industry. The Traeye (or Triay) family trace their lineage to African and Spanish ancestors who lived in St Augustine in the early nineteenth century. One of the earliest families to settle in Old Town Fernandina, they quickly became influential

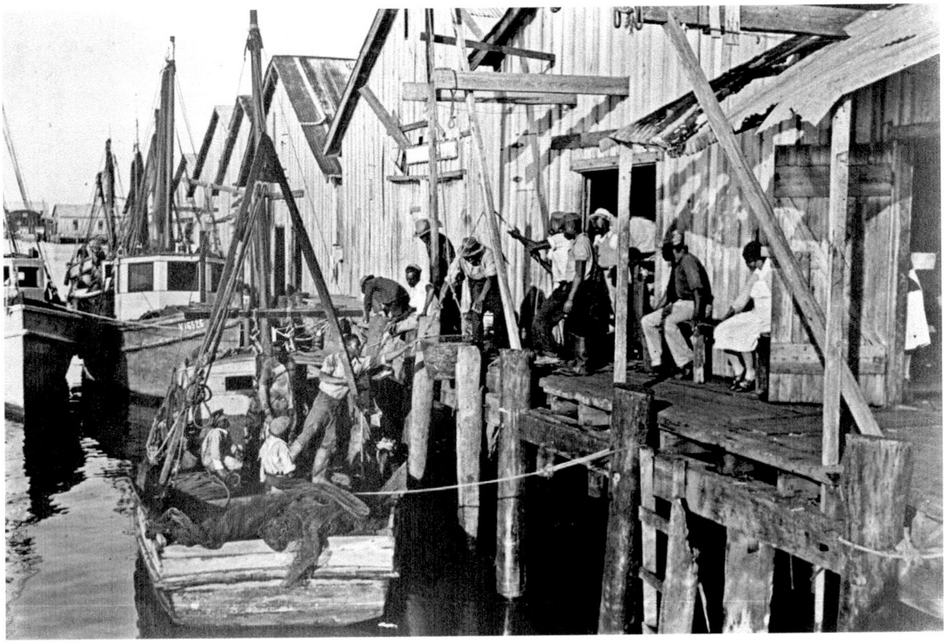

Unloading the catch from a shrimp boat. Shrimping provided sources of income for a large part of the community. *AIMH Archives courtesy of Billy Burbank III*

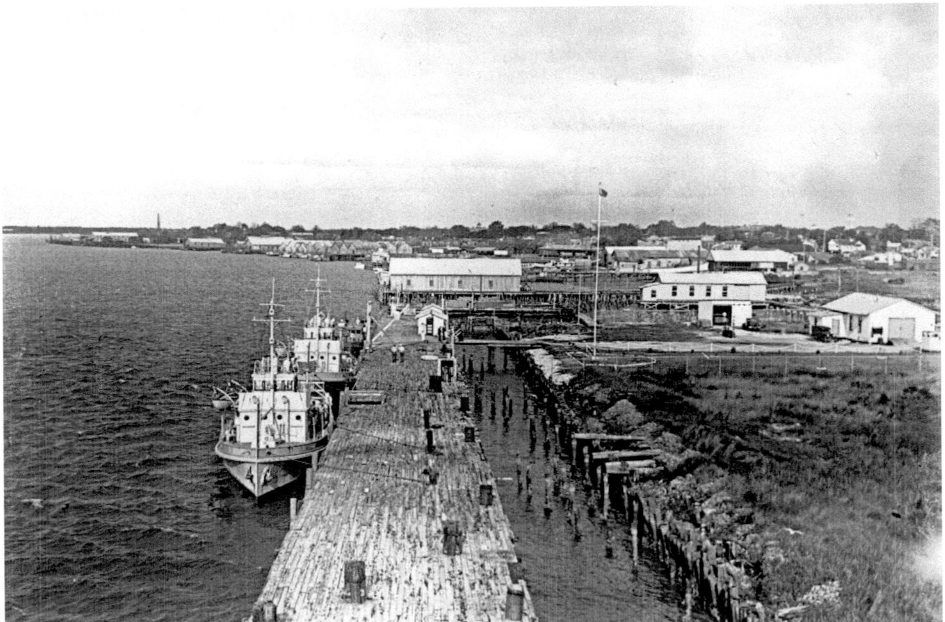

Old Town in the 1920s. *AIMH Archives*

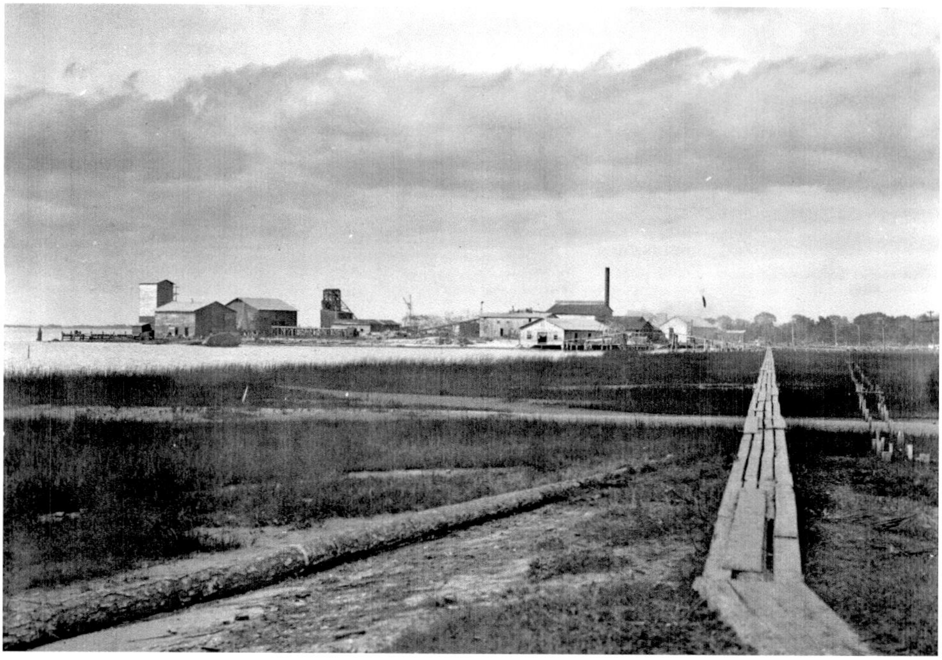

View of the Pogy Plant from Old Town. *AIMH Archives*

within the fishing community and maintained strong ties to the Catholic Church. Traeye descendants still reside on Amelia Island and in nearby Jacksonville.

Helen Traeye Frederick capsulized the experience of growing up in Old Town to the Fernandina News-leader:

> I grew up in Old Town, where everybody knew everybody else and we were all one big community family. My father, Herman Traeye, was a shrimper, and if he didn't catch any shrimp there were none to be caught. We children missed him when he was at sea for two or three days, but when the boat headed back, he would start blasting the foghorn as they approached the fort, and we knew Daddy was on his way home with the shrimp. We would run to meet him at the docks, and Captain Moss, Ed Livingston's uncle, would add to our pleasure by giving nickels and dimes to us children.

William H. McCollough (known as "Capt. Mac") was a Fernandina native and veteran shrimper who worked in local fishing for a number of years, on boats owned by Sloan Peterson, Harry Sahlman, Henry Lannon, and Joe Tringali. On Sunday, May 4, 1969, McCollough became the first and only African American captain to enter and win the Isle of Eight Flags Shrimp Boat Races, when he took second place in the 170-hp category racing one of Joe Tringali's boats, *The Ebb Tide*.

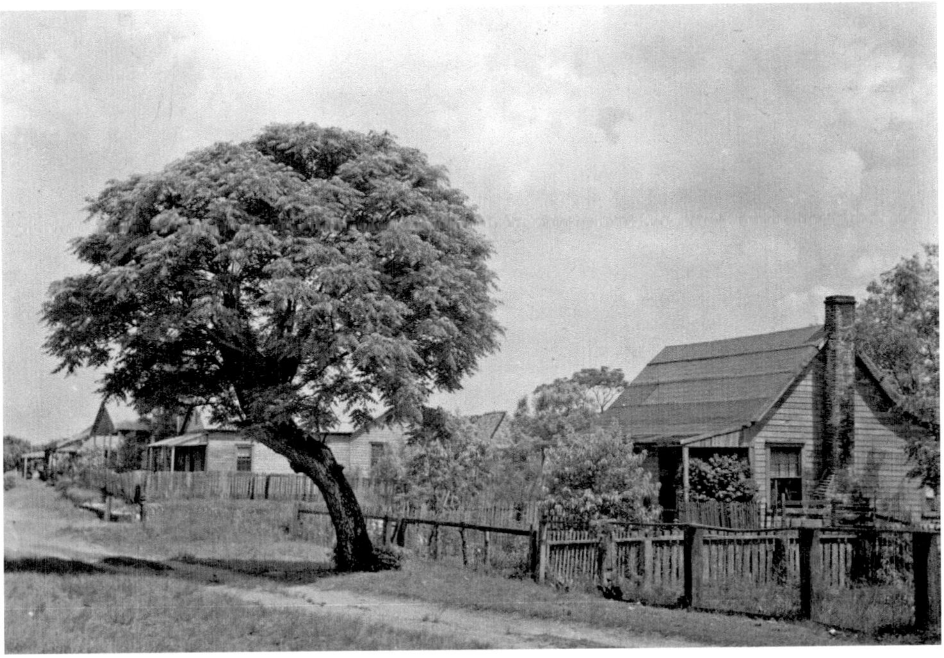

Above: Old Town Fernandina in the 1930s, photo taken from Someruelus Street. *AIMH Archives*

Right: Captain Wendell Herbert McCollough (1914-1993).

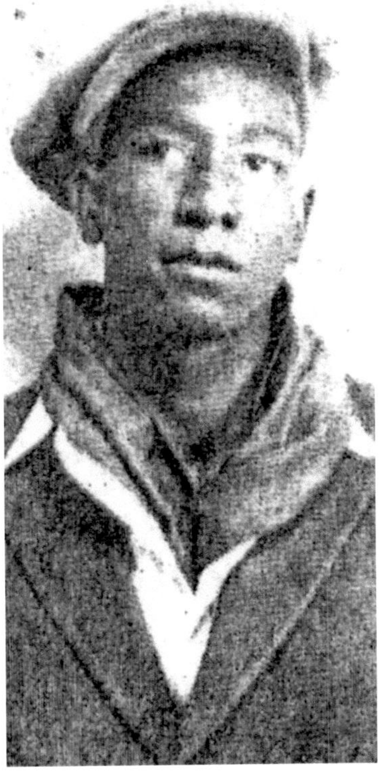

Captain Samuel Williams was a community elder, who operated what his son James Williams III remembers as "one of the slowest, worst looking" fishing boats in the area – the *W. B. Blades*. Despite his vessel's limitations and a lack of formal education, Williams was a capable seaman and fair captain, who kept his twenty-two man crew (consisting mostly of local African American men) well paid. His crew was able to fish commercially five to six months out of the year, using the rest of their time to mend nets or shrimp. Both on his boat and in the community, his son recalls that Williams "led by example." When Captain Williams spoke, local fisherman and other captains would draw close and listen with deference.[9]

Another predominately Afro-American community with ties to local fishing also existed just off the island, but still along the Nassau River, in the area known as Nassauville – where the entrepreneur Saul Goffin operated a processing plant. Goffin was a Russian immigrant who purchased the Nassauville cannery in 1893. He updated and expanded the facilities, constructing a compound for his family and the plant's workers (most of whom were African American), which came to be called "Goffinsville." The plant mainly processed shrimp, crab, and oysters. A commissary was built in 1899, along with a post office, and a school (a one room schoolhouse which taught students from first to eighth grade). The plant's workers built houses lining the road to the factory, and established their own church. A man named Prince Albert was one of the founders of the church, as well as one of its first preachers. By the early 1900s Goffinsville had basically become a self-contained community reliant on the Nassau River.

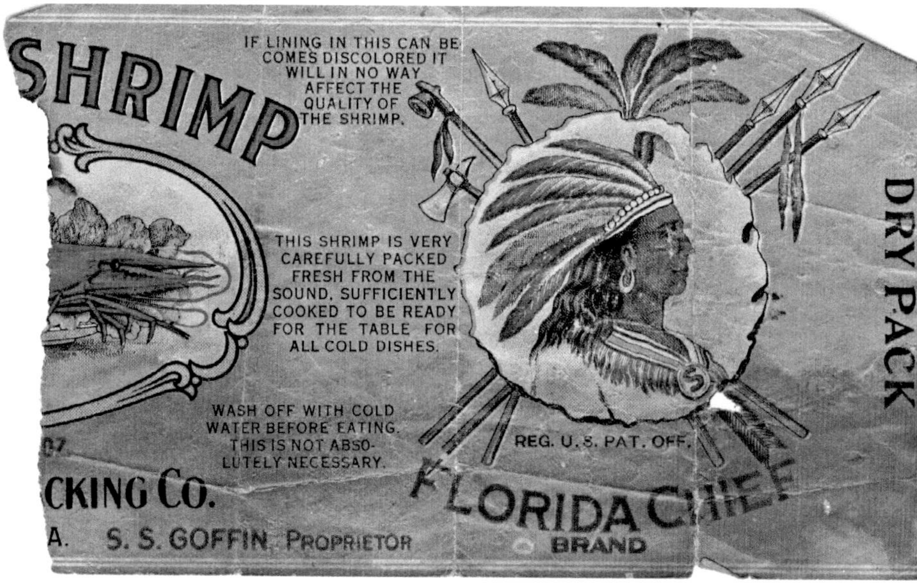

Label from the Florida Chief Brand Packing Co, owned by Saul Goffin. *AIMH Archives*

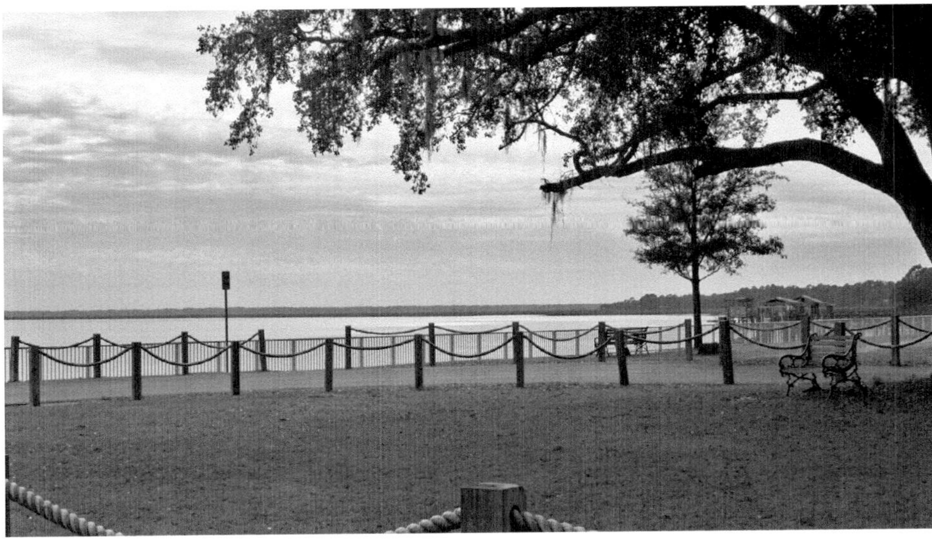

The Goffinsville Nassau River Park in 2014. This 19 acre park with a playground, nature trails, and a fishing pier commemorates the community of Goffinsville.

Saul Goffin passed away in the 1950s, and the factory's new owners failed to maintain the business. Much of the community that had built up around the compound moved to the island or elsewhere in search of opportunity. Erosion caused the remnants of the factory to the fall into the river within a few decades of the plant's closing. The old African American Church was reincarnated many years later, when Prince Albert's son Charles Albert Sr. donated land for a new building, dubbed Prince Chapel AME. Prince's grandson Charles Albert Jr. became a well-respected local educator and went on to serve terms on Fernandina's City Council and as the city's mayor.

Since shrimp boats operated with relatively small crews, Afro-American fishermen often found more consistent work in pogy fishing. Pogy crews were chosen by the captain of a particular vessel, and it was in the captain's best interest to have the best man for the job, regardless of skin color. Pogy crews were often either integrated, or made up exclusively of African Americans. Veteran pogy fisherman Neal Frink was the first African American captain on the east coast to get a license to operate a pogy boat.

Crews that were all African American, or led by black captains had to deal with discrimination at segregated ports as they followed the pogy fish up and down the east coast, some of which refused to even let them dock their boats. In an interview for the Amelia Island Museum of History, Neil Frink (son of Neal Frink) related stories former crew members told him of his father's canny solutions when faced with racism. Sometimes when steering his ship into segregated ports, he would be coarsely informed by the dock master that his boat would not be allowed to dock. Frink would feign that he was hard of hearing and stall the dock master long enough for his men to tie up the boat.

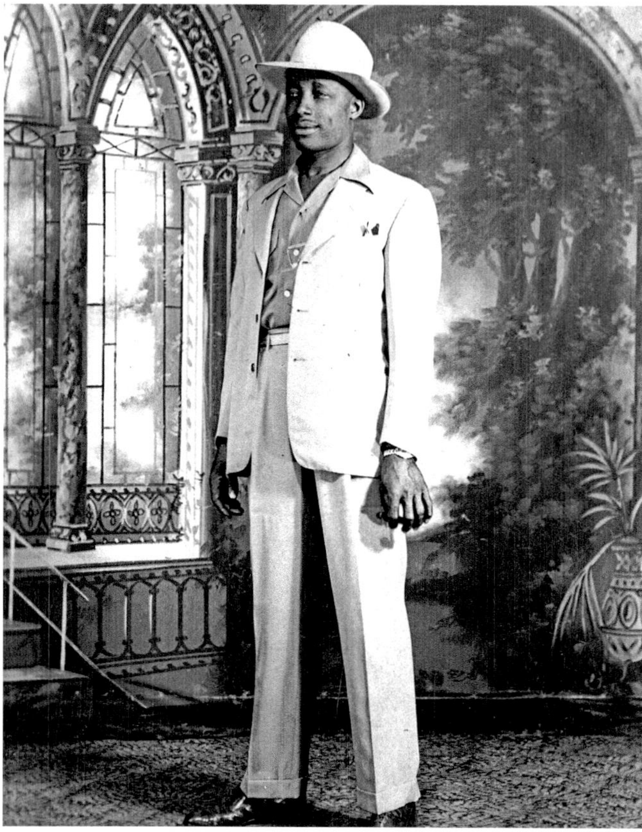

Neal Frink Sr. *AIMH Archives*

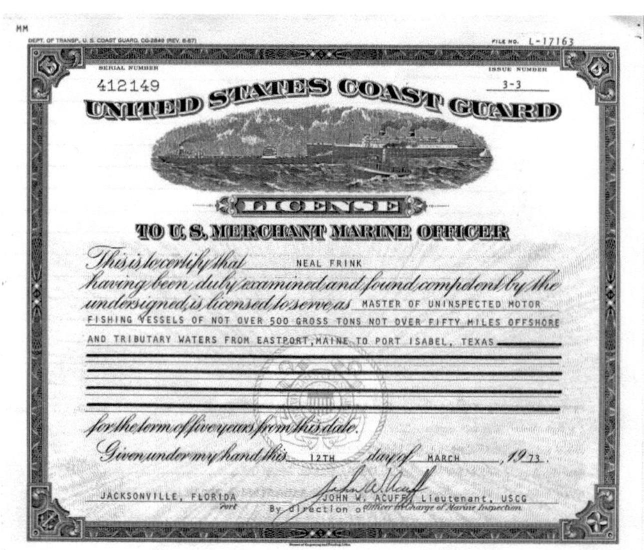

MM

DEPT. OF TRANSP., U. S. COAST GUARD, CG-2849 (REV. 6-67) FILE NO. L-17163

SERIAL NUMBER 412149

ISSUE NUMBER 3-3

UNITED STATES COAST GUARD

LICENSE

TO U. S. MERCHANT MARINE OFFICER

This is to certify that NEAL FRINK having been duly examined and found competent by the undersigned, is licensed to serve as MASTER OF UNINSPECTED MOTOR FISHING VESSELS OF NOT OVER 500 GROSS TONS NOT OVER FIFTY MILES OFFSHORE AND TRIBUTARY WATERS FROM EASTPORT, MAINE TO PORT ISABEL, TEXAS

for the term of five years from this date.

Given under my hand this 12TH day of MARCH, 1973.

JACKSONVILLE, FLORIDA
Port

JOHN W. ACUFF, Lieutenant, USCG
By direction of Officer in Charge of Marine Inspection.

Neal Frink's License to operate a pogy boat. *AIMH Archives*

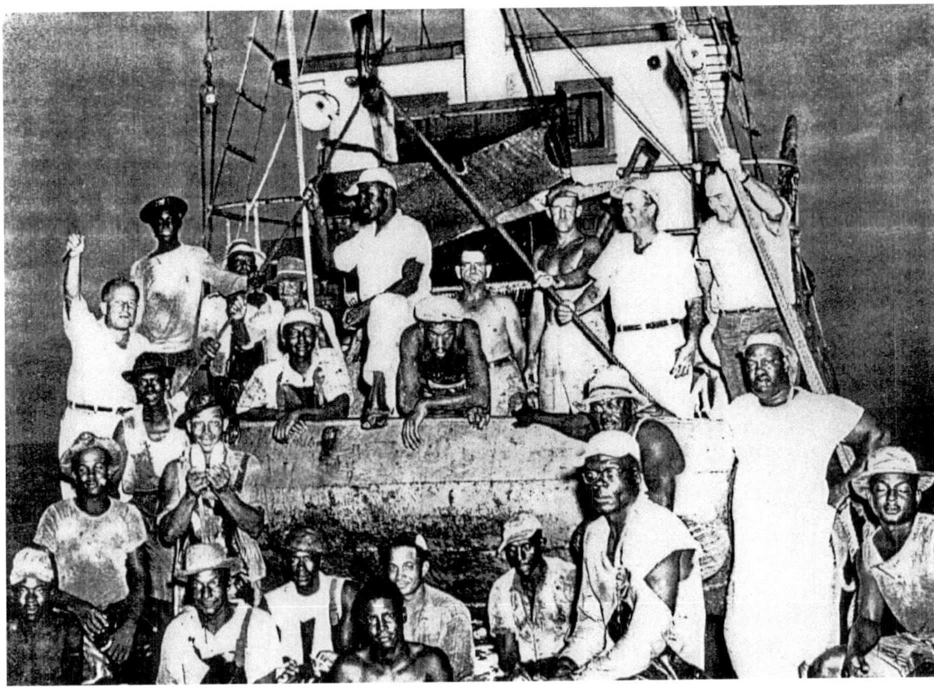

The crew of a pogy boat (possibly the *Tar Heel*). *AIMH Archives*

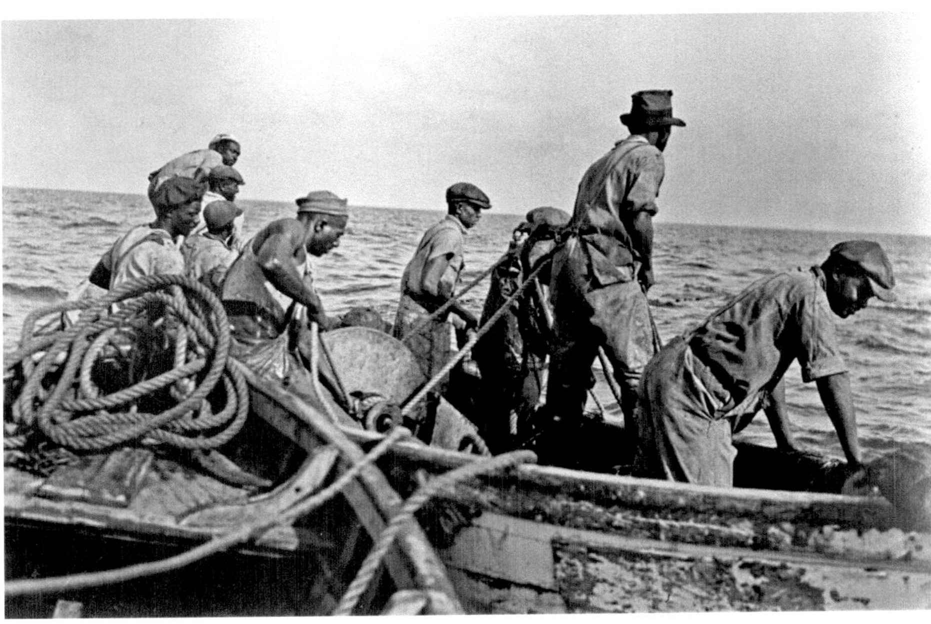

Pogy fishermen working the nets. *AIMH Archives*

Other times a light-skinned crew member would temporarily pose as captain, wearing Frink's captain's hat, and steer the ship into segregated ports. They would unload fish, get supplies, and as soon as they were clear of the dock Frink would resume his place at the wheel and send the man back to work. Neal Frink Sr. was also a highly regarded singer of chanties – one of the richest parts of the pogy fishing culture.

Menhaden Chanteys – songs the fishermen sang as they worked – share roots combining African-American work songs, spirituals, and sea shanties. They serve not only to pass the time and raise spirits, but the rhythm of the song can be used by the workers to synchronize their movements when hauling heavy nets. Many of the songs are call and response, with a leader (or "caller") offering up a line, answered by the rest of the men in harmony. There are standard verses, but in the work song tradition, there is a great deal of room for improvisation by the caller. Several of these songs have gospel origins like "Swing Low, Sweet Chariot," "My Way Seems So Hard," and "Remember Me."

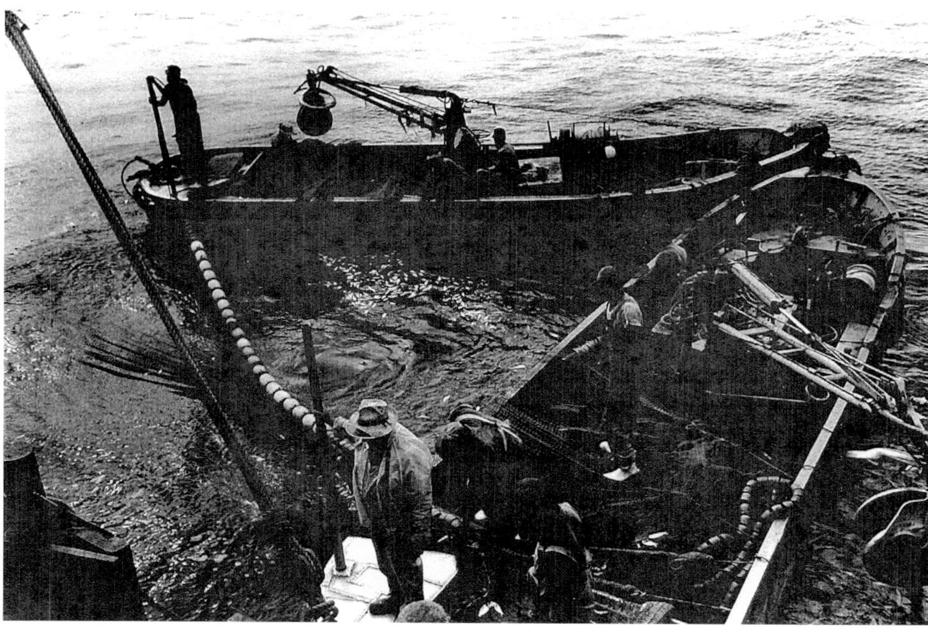

A pair of purse boats with their catch, c. 1950. *AIMH Archives, photo courtesy of Charles Corbett*

10

The Hardee Brothers: Outfitting an Industry

With all of the technical innovations and opportunities opening up in Fernandina for shrimpers, someone naturally had to step up to keep them outfitted in ropes, nets, and everything else needed to run a boat. This need would be filled by a family of brothers named Noble, John, and Ira Hardee who started a business in Fernandina which would eventually become a large and multi-faced enterprise, as well as one of the most important outfitters of shrimping equipment in the entire industry.

The boys' father, Dr. Charles Hardee, served as a surgeon in the 3rd Florida Infantry, later serving in Tennessee under his famous uncle Confederate General William Joseph Hardee (1815-1873). William J. Hardee (who earned the nickname "Old Reliable") was a West Point graduate from Camden County, Georgia, who fought for the US Army in the Second Seminole War and the Mexican-American War. In 1855 he published *Rife and Light Infantry Tactics for the Exercise and Manoeuvres of Troops When Acting as Light Infantry or Rifle* (more commonly known as *Hardee's Tactics),* which was the most popular drill manual used by both sides during the Civil War. Hardee joined the Confederate Army after his home state of Georgia seceded from the Union, and eventually reached the rank of Lieutenant General.

After the war, Dr. Hardee returned to his boyhood home of St. Mary's, Georgia (which is close to Fernandina and shares a place at the mouth of the St. Marys River). He married Mary G. Richardson of Duval County, Florida, and the couple had seven surviving children, who included John Richardson Hardee (born 1870), Noble Anthony Hardee (born 1879), and Ira William Hardee (born 1881). As a young man John Hardee (known to friends as J. R. or Capt. John) went to work for L. A. Davis at his saw mill in Crandal, Florida, and also earned money as a stevedore and fisherman. After an education in the seminary, younger brother Noble turned his attention to business. Together, the brothers opened a ship's chandlery (a retail business which specializes in equipment for ships), grocery and hardware store in Fernandina in the year 1900. By 1903 Ira Hardee had joined his elder brothers in business.

The store was originally located at the foot of Centre Street, in what was known as the Miller Building, but moved to the corner of Centre Street and 3rd in 1915. Catering to the needs of the then nascent shrimping boom was a natural fit, and the Hardee family's fortunes would expand along with the industry. The Hardees were the first retail providers

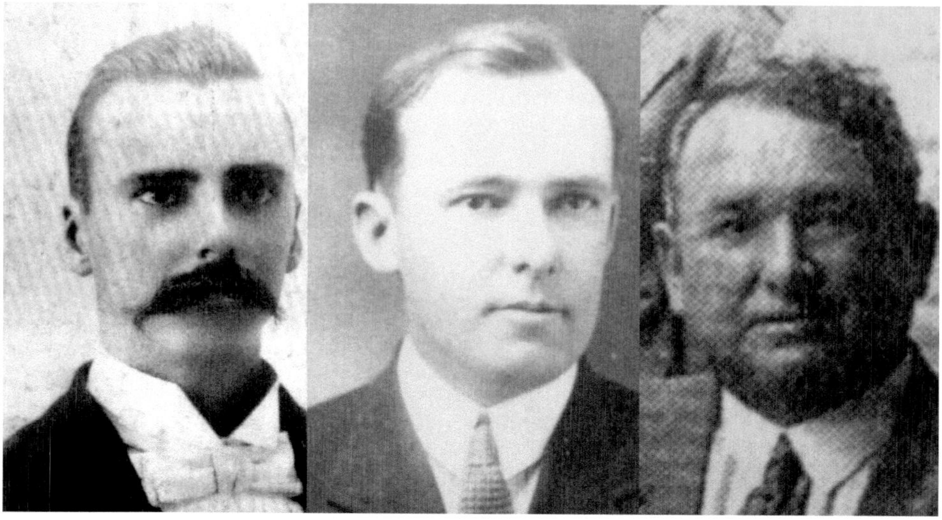

John R., Ira W., and Noble A. Hardee. *AIMH Archives*

to market a trawl made specifically for shrimping. Around 1915, the Hardee's enlisted the services of local net maker William Burbank I as a subcontractor. The agreement between the Hardees and Burbanks, consummated with just a handshake, would last almost eighty years.

Their success bred expansion, and all three brothers would pursue their individual interests. However, the brothers' diverging interests would in many ways remain linked, like separate fingers on the same hand. The three brothers partnered to open the South Atlantic Canning Company in 1916, with locations in Fernandina and St. Marys, Georgia (managed by another brother, Charles Hardee). In 1918, younger brother Ira bought his brothers out and took over the retail portion of the hardware store. The business was incorporated as the Standard Hardware Company, with Ira as its president, John as vice-president, and Noble later taking on the role of secretary-treasurer.

Noble focused his attentions on real estate and wholesale oil. J. R. preoccupied himself with the local shrimping industry and civic endeavors. He owned fish houses in Fernandina and St. Augustine, and served as the Port of Fernandina's Harbor Master, and on the Nassau County School Board. With his son John Jr. (a young, but able shrimper) J. R. started J. R. Hardee and Sons, a commercial shrimping operation with an eight boat fleet. Ira concentrated on the ambitious expansion of Standard Hardware. A keen and forward thinking businessman, Ira maintained a reputation for integrity and fair dealing among the citizens of Fernandina.

The company was successful in its early years, and as shrimping thrived the Hardees naturally sold more and more marine equipment to keep up with the increasing demand from shrimpers. The hard times that hit shrimpers during the Great Depression were acutely felt by Standard Hardware. The company was able to weather the depression, along with the general shift of the shrimping industry south and to the Gulf Coast.

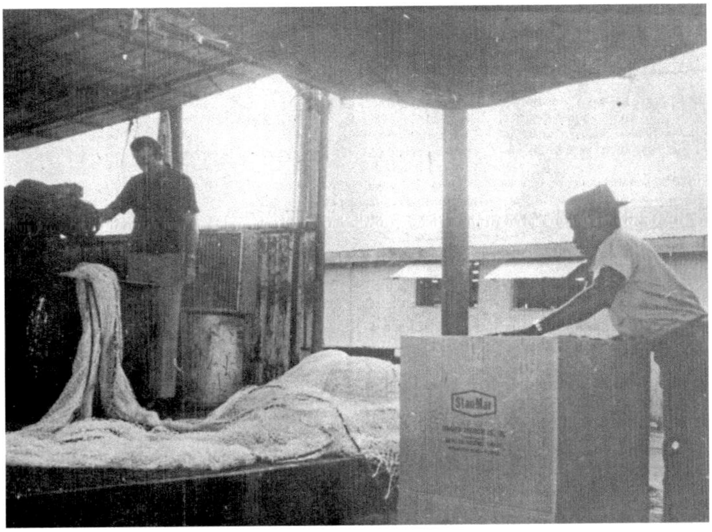

Dipping nets at Standard Marine. *AIMH Archives*

In 1948 the company lost one of its original leaders, Ira Hardee Sr., and a year later, a fire devastated the building on Centre Street that had been the company's home for nearly half a century. Ira's sons, Ira Hardee Jr. (called Bill), James Baker Hardee (called Baker), and Archibald Hinton Hardee (called Arch) took control of the company. At that point, the marine based portion of the enterprise, rechristened as Standard Marine Supply Company, relocated to the Huot Building on the corner of Alachua Street and Second Street. Constructed in 1882 by Frenchman C. H. Huot (a fire in 1878 destroyed the original building), the two-story, Commercial Vernacular building had served as a General Store and Saw Mill. The retail hardware operation remained at the Centre Street location, and took the name Hardee Brothers Company, Inc.

Like many other ventures born in Fernandina, Standard Marine (branded as STANMAR) had to follow the shrimp wherever they could be found. When the Tortuga shrimping beds were found in 1950, the Hardees established a presence there, outfitting the hundreds of boats who came with the boom. Two years later, the run at Campeche moved the center of shrimping yet again, this time to Tampa. The Hardees followed, opening the extremely successful Standard Marine Supply Company, which grew into one of the most successful commercial fishing suppliers in the southeast.

The Tampa office became the headquarters and the office of Baker Hardee (later succeeded by his son Jim), while Bill and Arch remained in Fernandina. After Bill passed away in 1970, Arch and Bill's wife Suzanne Davis Hardee both served as Standard Hardware's vice presidents until their retirement. The Standard Marine Supply Corporation had been born and grown with the shrimping industry in the early 1900s, but as industry withered in the last two decades of the nineteenth century, the writing was on the wall for STANMAR. The locally born company, which had seen the rise and fall of modern shrimping, dissolved in 2008.

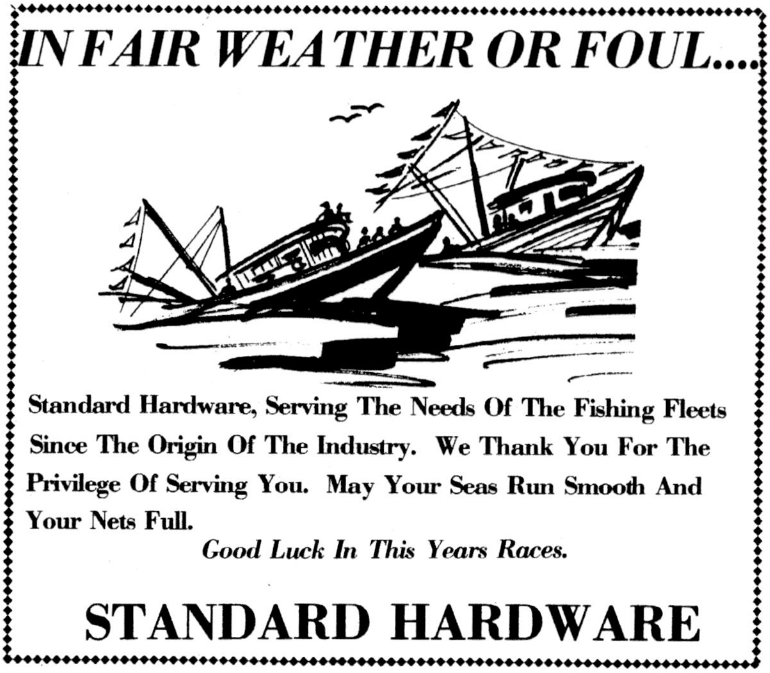

Standard Hardware advertisement from the early 1970s. *AIMH Archives*

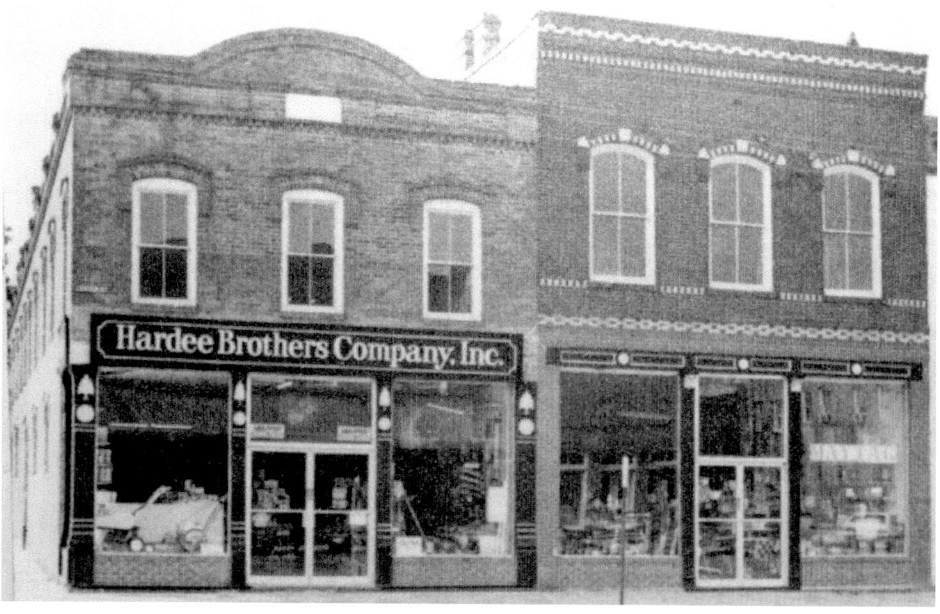

Hardee Brothers at the corner of 3rd St and Centre Street. This was the company's headquarters from 1915 until 1949. *AIMH Archives*

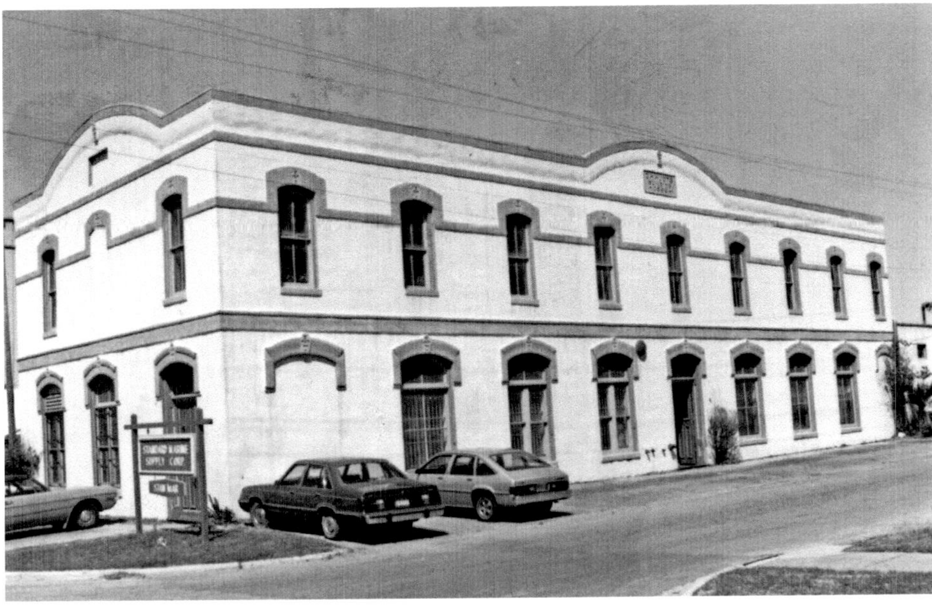

The Huot Building on North 2nd Street – Standard Marine's office in Fernandina. *State Archives of Florida; photo by Eric Larsen*

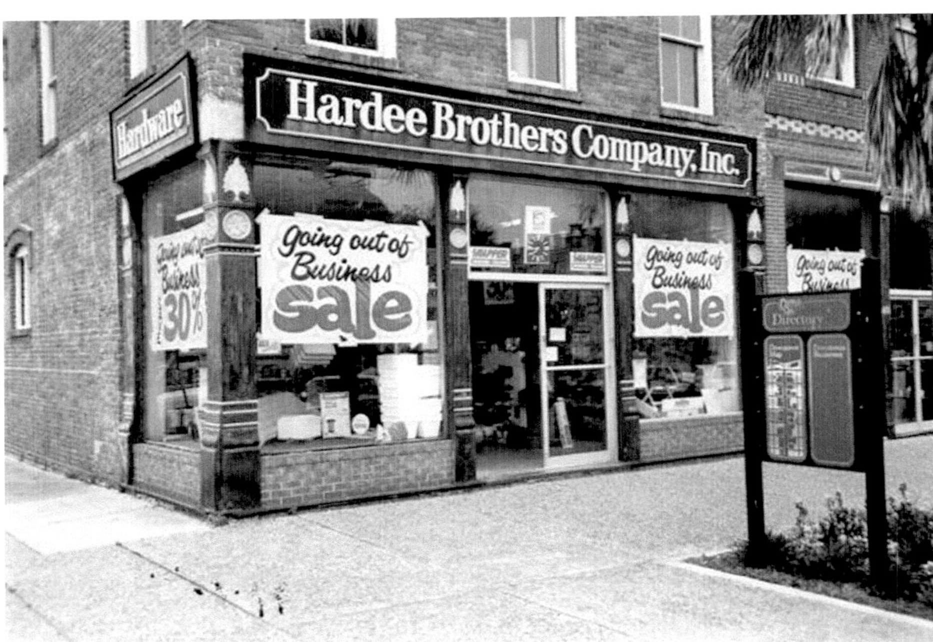

Standard Hardware just before the closing of the Centre Street location. At that point the hardware and marine supply operations were recombined. *AIMH Archives*

11

The Shrimp Festival

It is a long standing tradition among shrimpers to gather together each year to celebrate the start of the season, and to pray for a safe and lucrative year. In 1963 the city of Fernandina began looking to turn this yearly ritual into a celebration similar to Tampa's Gasparilla Festival. The aim was to bring tourists to Amelia Island, and make use of the city's beautiful harbor and historic downtown area. Early in the festival's history it was referred to as the "Shrimp Boat Festival," which eventually became simply, the "Shrimp Festival."

Many of the festival's early traditions can be traced back to a celebration begun by the Lion's Club in the 1950s, called the "Fiesta of Eight Flags." Held in June, this week long celebration included the opening of the beach ceremony, the Fernandina Open Invitational Golf Tournament, a historical pageant held at Fort Clinch called *The Romance of Eight Flags*, a dance, a beauty pageant (the forerunner of today's Miss Shrimp Festival Pageant), and a parade that stretched the width of the island. The Shrimp Boat Festival combined many of these festivities with a shrimp boat race. Attendants would be able to watch local captains pit their boats against each other for bragging rights (as well as a trophy and cash prizes). The first race was held on June 7, 1964; the event quickly became a crowd favorite.

In the race's first year (as well as the two that followed), Joe Tringali's *Dixie Queen* captured first place, and the $100 prize that went along with it. At first only local boats from Fernandina and Saint Marys entered, but by 1966 the race began attracting attention from national publications, and soon boats from other towns along the coastline began to join. People came from all over the region for this unique and exciting form of entertainment. Before the race, a grand procession was held where all of the boats in the competition would pass by the eager spectators. The boats would be heavily decorated, and owners would invite their friends and family to ride on board. Many locals have fond memories of riding on shrimp boats during the parade, and waving at the crowds lining Fernandina's docks downtown.

Companies like General Motors and Caterpillar sponsored the boats, and used the races as a venue to test their newest prototype engines. Shrimpers would spend the week before the race hot rodding their boats to move much faster than they would ever need to during a shrimping run. The race expanded to include different categories, including the up to '175 horsepower', the up to '250 horsepower', and the 'Unlimited' categories –

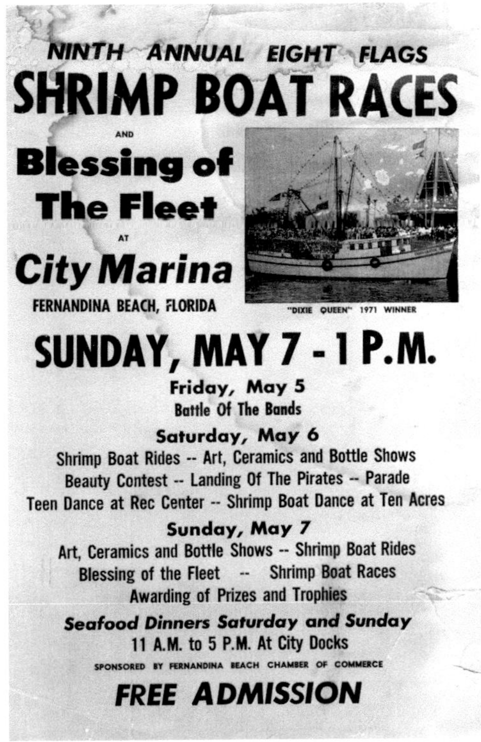

Right: Poster from the 9th Annual Isle of 8 Flags Shrimp Boat Races. *AIMH Archives*

Below: Young women in the Fiesta of Eight Flags Beauty Contest—*c.* 1950s. Miss Jacksonville was named that year's winner. *AIMH Archives*

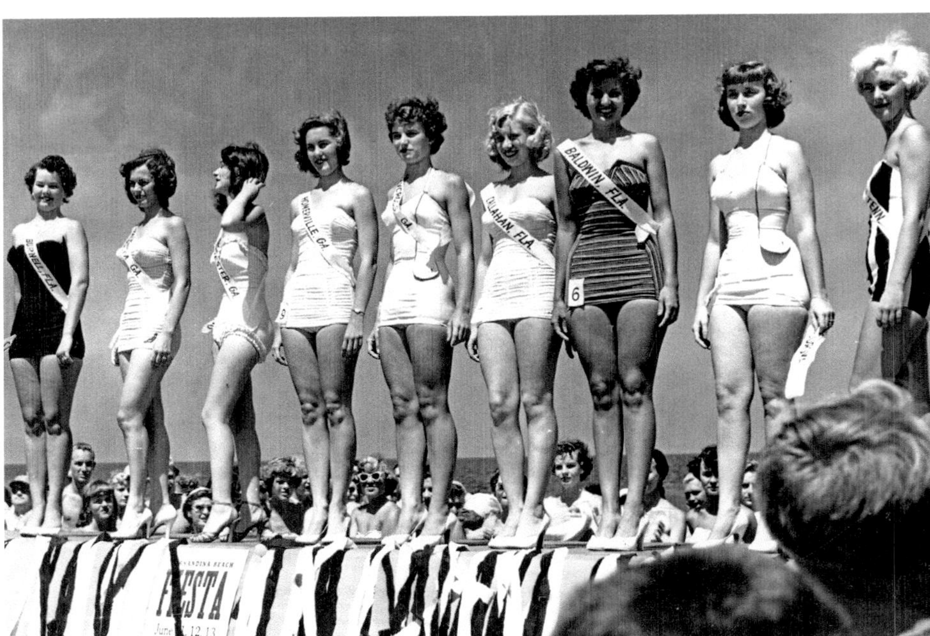

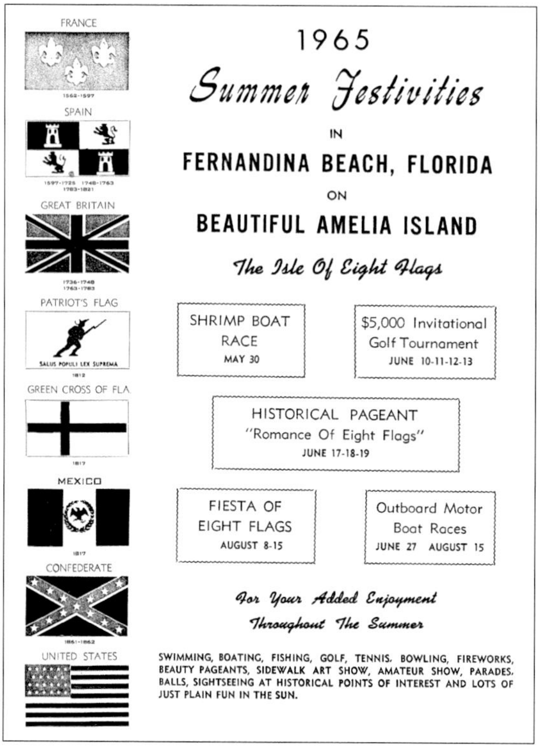

Left: Program from the 1965 Isle of Eight Flags Festival. *AIMH Archives*

Below: The Dixie Queen during the 1970 shrimp boat parade. *State of Florida Archives; photo by Karl E. Holland*

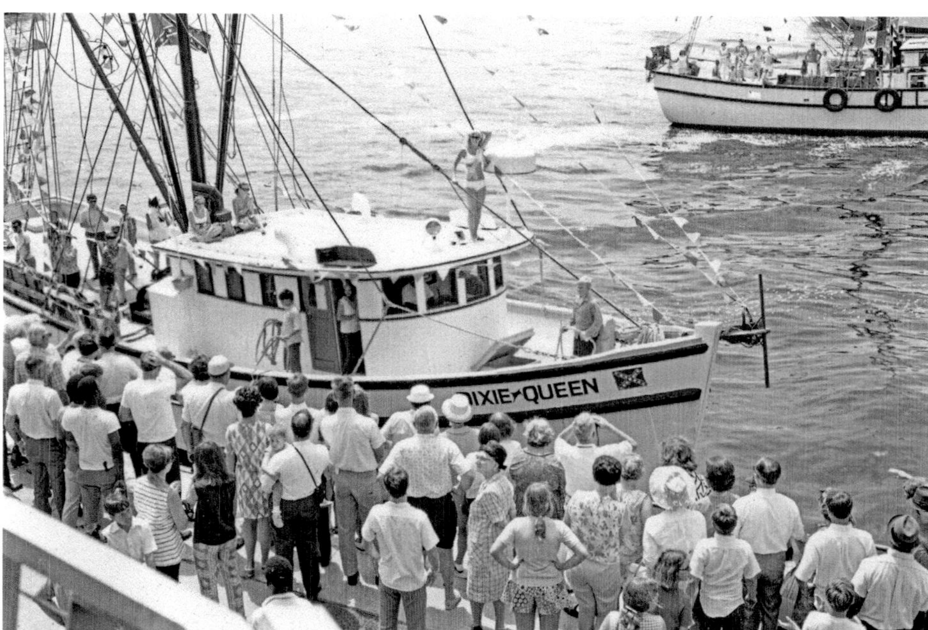

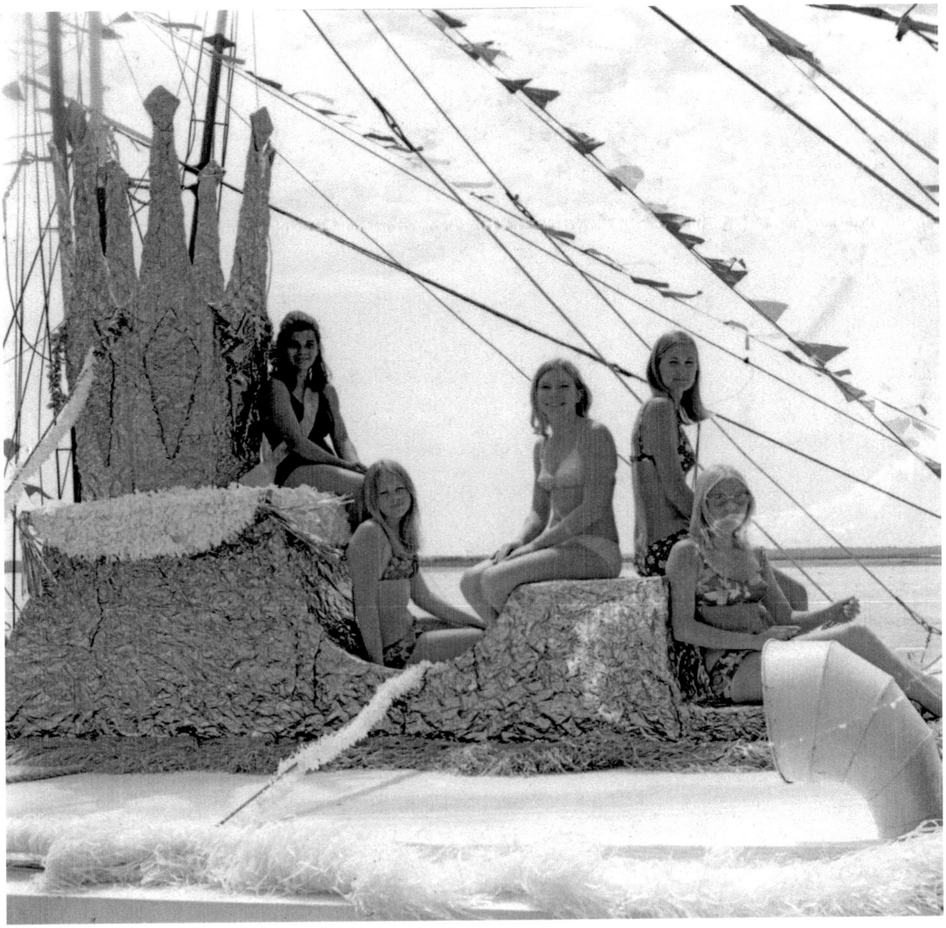

Miss Shrimp Fest contestants in the 1970 parade. State of Florida Archives. *Photo credit: Burrall*

opening up the field of competition. Awards were given out for "Best Decorated," and "Most Original Decoration." Originally, the boats raced on a straight course, racing out to the pier on South Fletcher Avenue and then doubling back. In 1968 the course was changed to a figure 8 so that the spectators could see more of the race.

The races had become so popular that for many years the weekend's festivities were known as "The Shrimp Boat Festival." However, by the mid-1970s the rising cost of fuel and the looming downturn of the shrimping industry had the made the races too expensive for boat owners and captains to justify their participation. 1976 would be the race's final year, and Captain Joe Tringali's *Dixie Queen* would once again emerge victorious with a first place win in the 'Unlimited' class. It was a fitting bookend to the twelve year history of Isle of Eight Flags Shrimp Boat Races, the event which put today's Shrimp Festival on the map.

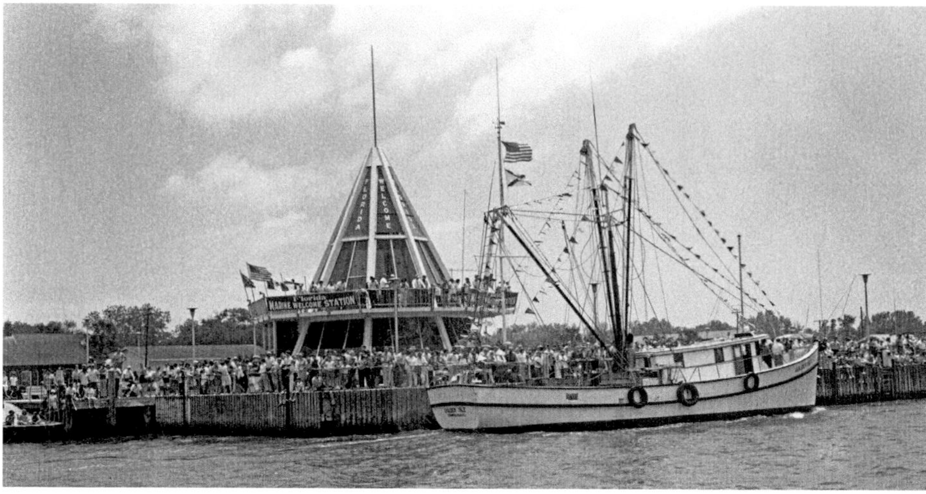

The 1969 Shrimp Boat Races. State Archives of Florida. *Photo by Karl E. Holland*

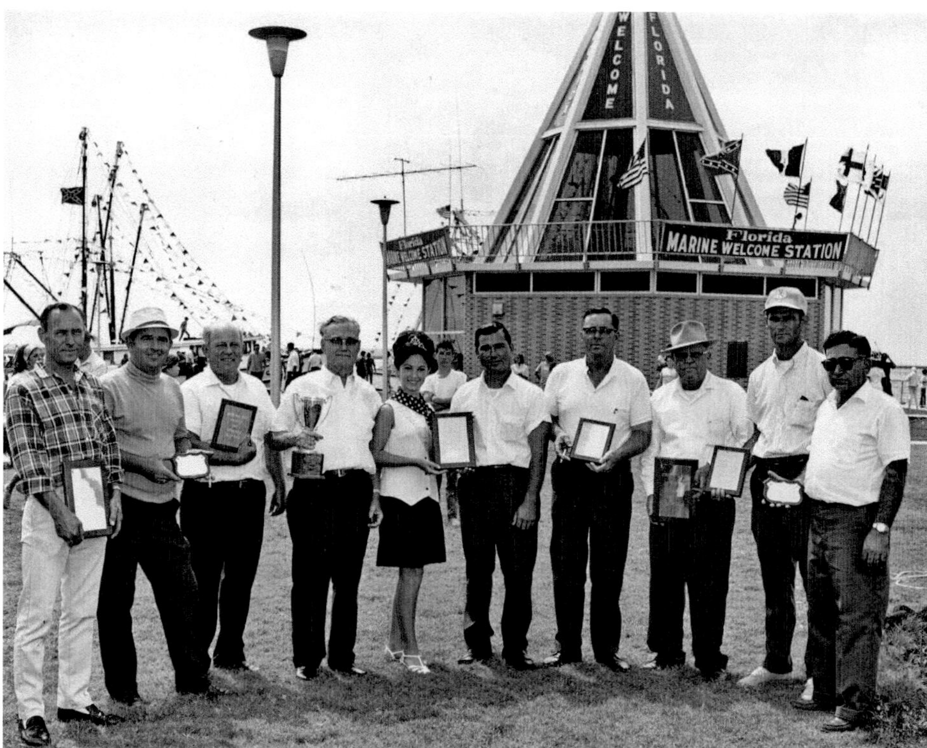

Shrimp Boat Race winners pose with festival queen. *From Left to right:* Earl Cook, Robert Lannon, Alvin Anderson, Sloan Peterson, Shrimp Boat Festival Queen Dixie Massey, Bunny Sterling, H. A. Cook, Jim Brazel, Alvin Dickley Jr., and Papa Joe Tringali. Fernandina Beach News-leader.

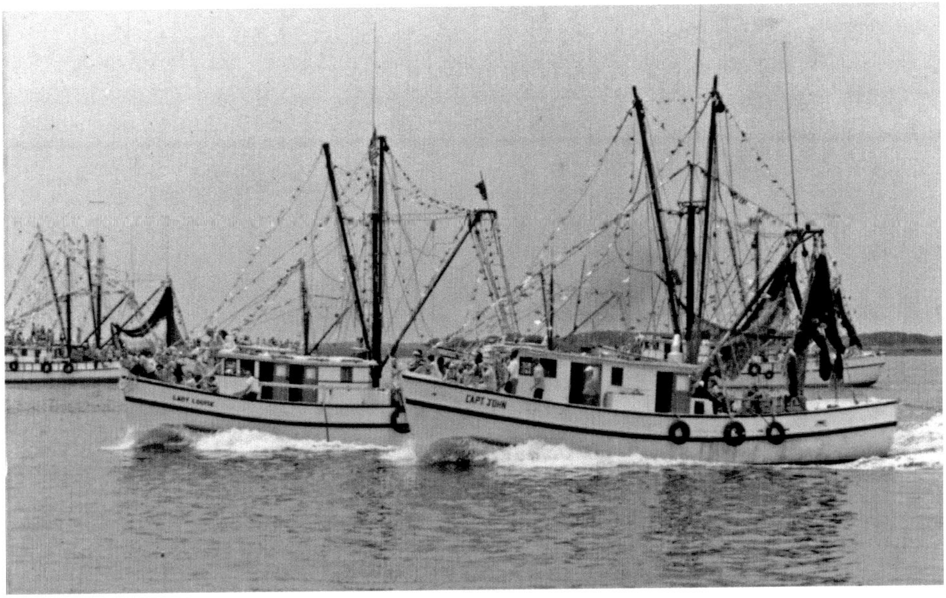

Shrimp Boats racing at Fernandina in the 1970s. *AIMH Archives*

One tradition that began early in the festival's history and has remained a part of the celebration to this day is known as "The Blessing of the Fleet." This is a ceremony in which a priest stands on the docks and asks God's blessing for the shrimp boats as they pass him. The custom was probably brought to Amelia Island from the Mediterranean by Italian and Greek fishermen, who were predominantly Catholic. The ceremony was of great importance to shrimpers, as their livelihood depended on nature's cooperation.

By 1971, attendance had increased to around 15,000 and the parade attracted the largest crowd ever up to that time. That year saw the first Landing of the Pirates, which occurred on Saturday at noon, and was directly preceded by the judging of the pirate's beards (which was sponsored by The Marina restaurant). That same year, the first professionally judged art show was held at the Chamber of Commerce. This set the stage for the high quality of art the festival would come to be known for. By 1972 the art show had grown to include 130 exhibitors. The festival's emphasis began to shift away from the shrimp boat races, and focus on arts and crafts. As the years passed the festival's art section expanded, and began to line Centre Street, attracting appreciators of art from all over the world. In 1973 the festival took on the pattern it would maintain for the next several years. The Landing of the Pirates would occur on Friday night to open the festival, with the Miss Shrimp Pageant moving to Saturday, with the art show, and beard contest.

Also in 1973, a group began that would become integral to the festival's atmosphere – the Fernandina Pirate Club. The Pirates Club is a benevolent organization that adds flair to the festivities by dressing as pirates. They kick off the festival on Friday night by landing

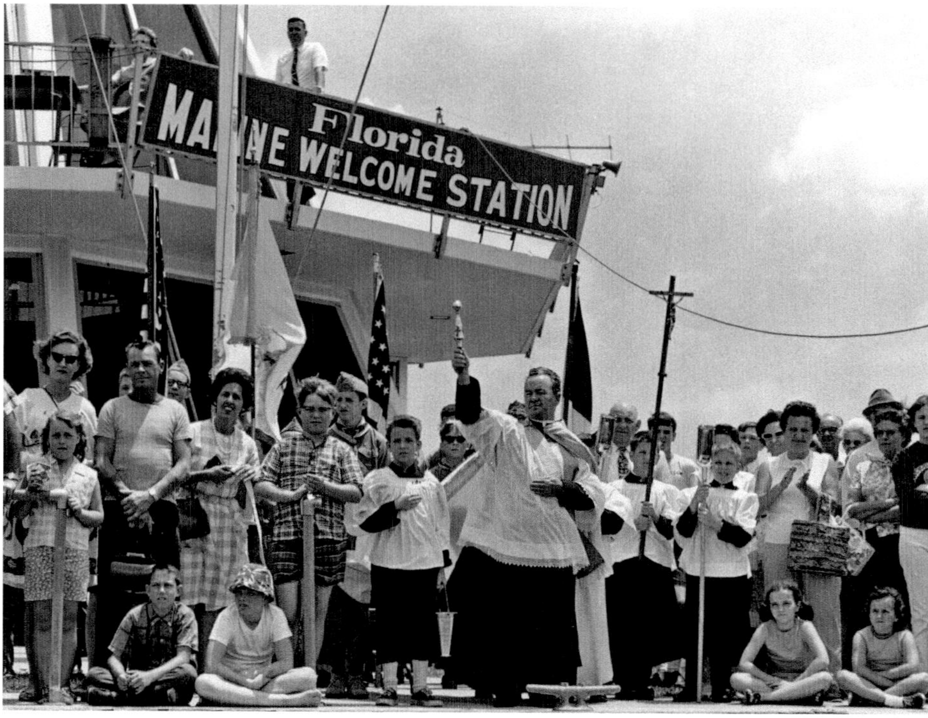

Father Duncan blessing the fleet at Fernandina. *AIMH Archives*

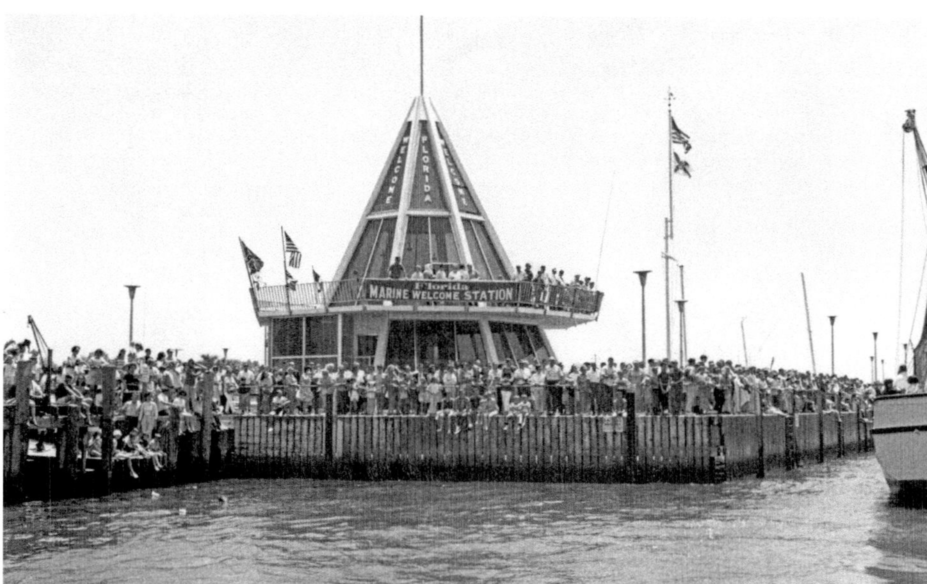

Spectators watching the 1972 Blessing of the Fleet. State Archives of Florida. *Photo by Karl E. Holland*

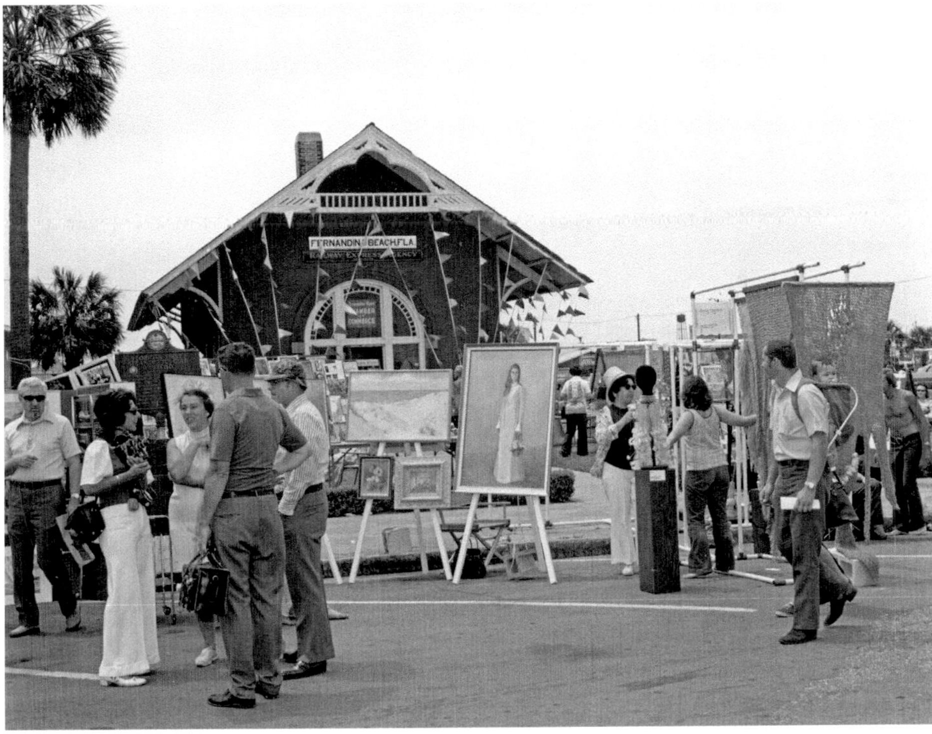

Art sold on Centre Street at the 1973 Shrimp Festival. *State Archives of Florida; photo by Karl E. Holland*

at the harbor in downtown Fernandina and "storming" the port, firing off a cannon and pistols, and interacting with the crowd. In the 1970s the Pirates were famous for their parade float, a 'brig' (or jail cell) that they would fill with beardless festival attendees, asking for a ransom (charitable donation) to secure their release. Later, they replaced the brig with a large, rolling pirate ship, always a parade highlight. Today, the Pirates provide scholarships for local students and make appearances for community and charitable organizations across the region.

Today the festival is one of the largest in the southeastern United States. It attracts over 150,000 visitors over a two day period, and features over four hundred vendors selling art, crafts, and local food. Over the years, the festival has managed to hang on to some early traditions, like the Blessing of the Fleet, the parade, and the Miss Shrimp Festival Pageant, but it has also grown to include an array of other attractions like the fireworks show, the pirate contest, and the Kid's Fun Zone.

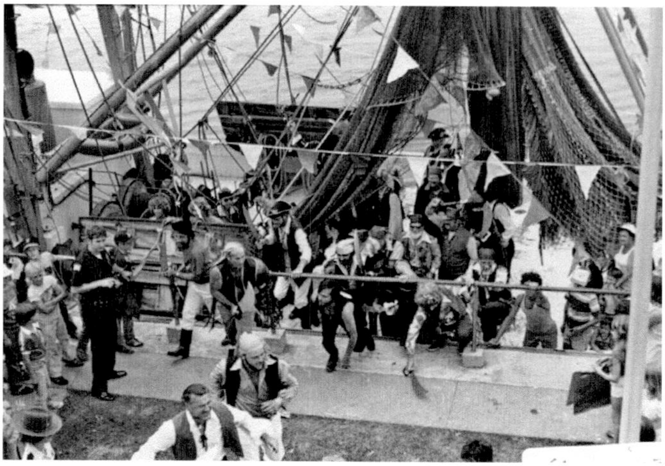

Left and below: An early Pirate Invasion. *AIMH Archives*

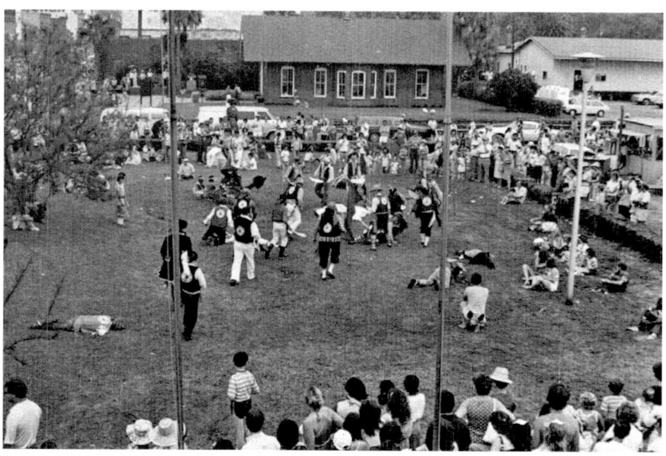

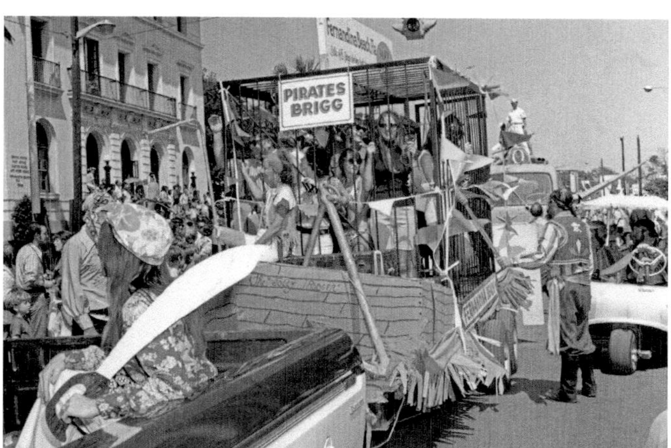

The Pirate's and their brig at the 1973 parade. *State Archives of Florida; photo by Karl E. Holland*

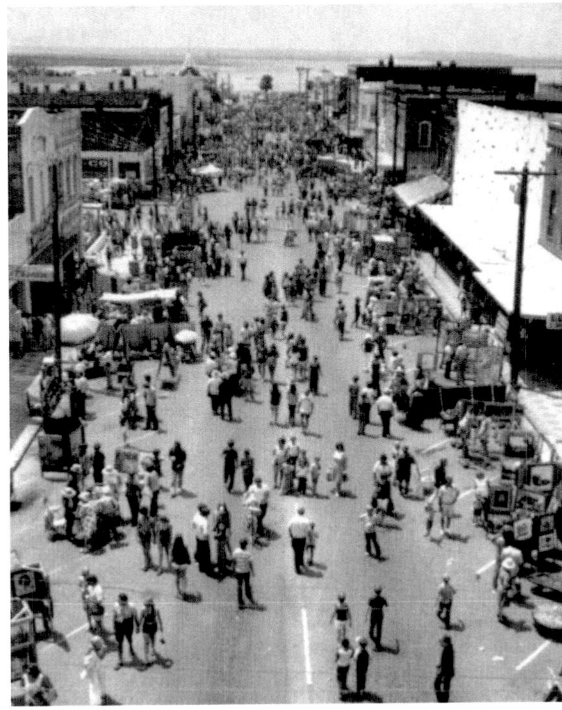

Centre Street during the 1974 festival. *AIMH Archives*

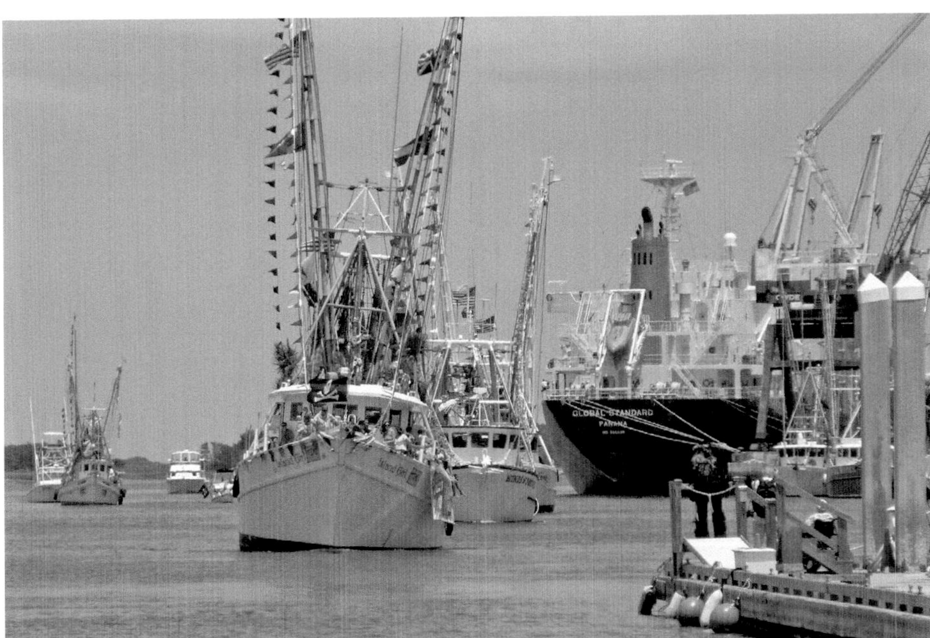

A modern Blessing of the Fleet. *AIMH Archives AIMH Archives*

SHRIMP BOAT RACE WINNERS
AND OTHER AWARDS FROM 1964 TO 1976

1964
1ST PLACE *Dixie Queen* captained by Joe Tringali
2ND PLACE *Florida Boys* captained by David Cook
3RD PLACE *The Haze* owned by David Cook

1965
1ST PLACE *Dixie Queen* captained by Joe Tringali
2ND PLACE *Delores D* captained by Earl Cook
3RD PLACE *Sea Bring* out of Saint Augustine

1966
Unlimited Class
1ST PLACE *Dixie Queen* captained by Joe Tringali
2ND PLACE *Chickadee* captained by Bunny Sterling
3RD PLACE *Florida Boys* captained by Ed Cook

Winners in Other Classes
Wanda and Peggy captained by Jr. Cook
Miss Nina Jo captained by Cliff Litties
Golden Isle captained by H. A. Cook

1967
Unlimited Class
1ST PLACE *Chickadee* captained by Bunny Sterling
2ND PLACE *Jet Stream* captained by Gus Jones
3RD PLACE the *Eight Flags* captained by Cecil Bennett

Up to 250 Horse Power Class
1ST PLACE *Florida Boys* captained by Ed Cook
2ND PLACE *Miss Nina Jo* captained by Cliff Litties
3RD PLACE *Delores D* captained by Earl Cook

Up to 175 Horse Power Class
1ST PLACE *Golden Isle* captained by H. A. Cook
2ND PLACE *Jabberwork* captained by Robert Lannon
3RD PLACE *Little David* captained by Hilton Cook

Other Awards

Best Decorated	*Golden Isle* captained by H. A. Cook
Most Original	*Miss Melody* captained by Earl Cook
Honorable mention	*Bossy Betty* captained by George Scott and *Jabberwork* captained by Robert Lannon,

1968

Unlimited Class

1st Place *Chickadee* captained by Bunny Sterling
2nd Place *Lady Belle* captained by Alvin Dickey
3rd Place N/A

Up to 250 Horsepower Class

1st Place *Gulf King* captained by Earl Cook
2nd Place *Golden Isles* captained by H. A. Cook
3rd Place N/A

Up to 175 Horsepower Class

1st Place *Little David* captained by Jim Brazell
2nd Place *Sunrise* captained by Alvin Anderson
3rd Place N/A

Other Awards

Best Decorated	*Dixie Queen* captained by Joe Tringali
Most Original Decoration	*Jabberwork* captained by Robert Lannon

1969

Unlimited Class

1st Place *Eight Flags* captained by Cecil Bennett
2nd Place *Lady Annie* captained by Jr. Cook
3rd Place *Chickadee* captained by Sloan Peterson

Up to 250 Horsepower Class

1st Place *Little David* captained by Hilton Cook
2nd Place *Golden Isle* captained by H. A. Cook

Up to 175 Horsepower Class

1st Place *Miss Libby* captained by George Scott
2nd Place *Ebb Tide* captained by William "Capt Mac" McCullough

Other Awards

Best Decorated	*Eight Flags* captained by Cecil Bennett
Most Original Decoration	*Bossy Betty* captained by Buddy Miller (owned by George Scott)

1970

Unlimited Class

1ST PLACE *Dixie Queen* captained by Joe Tringali
2ND PLACE *Lady Annie* captained by Jr. Cook
3RD PLACE N/A

Up to 250 Horsepower Class

1ST PLACE *Golden Isle* captained by A. B. Cook, Jr.
2ND PLACE *Little David* captained by D. A. Cook

Up to 175 Horsepower Class

1ST PLACE *Lady Louise* captained by Marion DuBose
2ND PLACE *Trade Winds* captained by E. E. Lewis

Other Awards

BEST DECORATED N/A
MOST ORIGINAL Decoration N/A

1971

Unlimited Class

1ST PLACE *Dixie Queen* captained by Joe Tringali
2ND PLACE N/A
3RD PLACE N/A

Up to 250 Horsepower Class

1ST PLACE *Little David* captained by Hilton Cook
2ND PLACE N/A

Up to 175 Horsepower Class

1ST PLACE *Lady Louise* captained by Marion DuBose
2ND PLACE N/A

Other Awards

BEST DECORATED *Chickadee* captained by Bunny Sterling
MOST ORIGINAL Decoration *Gulf King* captained by Earl Cook

1972

Unlimited Class

1ST PLACE *Virginia Gray* captained by Billy Ipock
2ND PLACE *Dixie Queen* captained by Joe Tringali
3RD PLACE N/A

Up to 250 Horsepower Class

1ST PLACE *Wanda and Peggy* captained by John Cook
2ND PLACE *Little David* captained by Hilton Cook

Up to 175 Horsepower Class
1st Place *Lady Louise* captained by Marion DuBose
2nd Place *Miss Libby* captained by Jack Ipock

Other Awards
Best Decorated *Golden Knight* captained by H. A. Cook
Most Original Decoration *Little David* captained by Hilton Cook

1973
Unlimited Class
1st Place Dixie Queen captained by Joe Tringali
2nd Place Lady Annie captained by Jr. Cook
3rd Place N/A

Up to 250 Horsepower Class
1st Place *Florida Boys* captained by Ed Cook
2nd Place *Miss Dayle*

Up to 175 Horsepower Class
1st Place *Lady Louise* captained by Marion DuBose
2nd Place *Captain John*

Other Awards
Best Decorated N/A
Most Original Decoration N/A

1974
Unlimited Class
1st Place *Miss Carol J* captained by DeWayne Phillips
2nd Place N/A
3rd Place N/A

Up to 250 Horsepower Class
1st Place *Golden Isle* captained by H. A. Cook
2nd Place N/A

Up to 175 Horsepower Class
1st Place *Miss Hazel* captained by Lewis Bennett
2nd Place N/A

Other Awards
Best Decorated N/A
Most Original Decoration N/A

1975

Unlimited Class
1st Place *Gray Ghost* captained by Dink Miner
2nd Place *Dixie Queen* captained by Joe Tringali
3rd Place N/A

Up to 250 Horsepower Class
1st Place *Little David* captained by Hilton Cook
2nd Place N/A

Up to 175 Horsepower Class
1st Place *American Beauty* captained by Pete Caldwell
2nd Place N/A

Other Awards
Best Decorated N/A
Most Original Decoration N/A

1976

Unlimited Class
1st Place *Dixie Queen* captained by Joe Tringali
2nd Place *Cindy*
3rd Place N/A

Up to 250 Horsepower Class
1st Place *Sarah Anne* captained by Cecil Bennett
2nd Place *Kathy captained* by Danny Bennett

Up to 175 Horsepower Class
1st Place *Lady Louise* captained by Marion DuBosev
2nd Place *American Beauty*

Other Awards
Best Decorated N/A
Most Original Decoration N/A

12

The End of an Era?

Even as many of the early shrimping pioneers left Fernandina in the 1920s and '30s, shrimping was still a thriving international industry, and it still continued to provide a sustainable living for many families in the area. Even in the 1950s and '60s, it was possible to see up to forty or fifty boats at a time tied up at Fernandina's docks. Local shrimpers could bring home $1,200 to $2,000 a week during the season. Charlie Cash's grocery store downtown catered to shrimpers and would open up at all hours to make deliveries to shrimp boats.

The business side of the industry has changed a great deal over the years. Local shrimpers and the workers in fields adjacent to the industry have had to weather the many transitions. In the 1940s both of Fernandina's remaining shrimp canneries closed down. By that time, shrimp were being put on ice to be shipped to Jacksonville for packaging and distribution. World War II was a challenging time for the shrimping industry on a national level, as fishermen had to circumvent rationing and other wartime measures. But as always Fernandina's shrimpers found a way to adapt and preserve. For example, local shrimper Manuel Carinhas found a canny way to get around mandatory radio blackouts by using carrier pigeons to keep his boats in communication with one another.

In the mid-1970s shrimpers were hit hard by a combination of factors, from which the local industry has never recovered. Today only a handful of shrimp boats sit docked in downtown Fernandina. Rising fuel prices have made the operation costs of running a boat untenable for many shrimpers. Increased government regulation on fisheries have also added expenses. Shrimpers must have a permit to operate a boat, which are heavily restricted, and fisheries are closely monitored by state and federal governments. In 1988 Congress issued a bill which requires that Turtle Exclusion Devices be installed in all trawl nets. The State of Florida passed legislation in 1995, which prohibited shrimping within a mile of shore.

The decline of local shrimping has been noticeable and steady. Shrimp boats used to be part of the landscape when looking out at Fernandina's waterways. Even residents who knew nothing about the ins and outs of shrimping had gotten accustomed to seeing them.

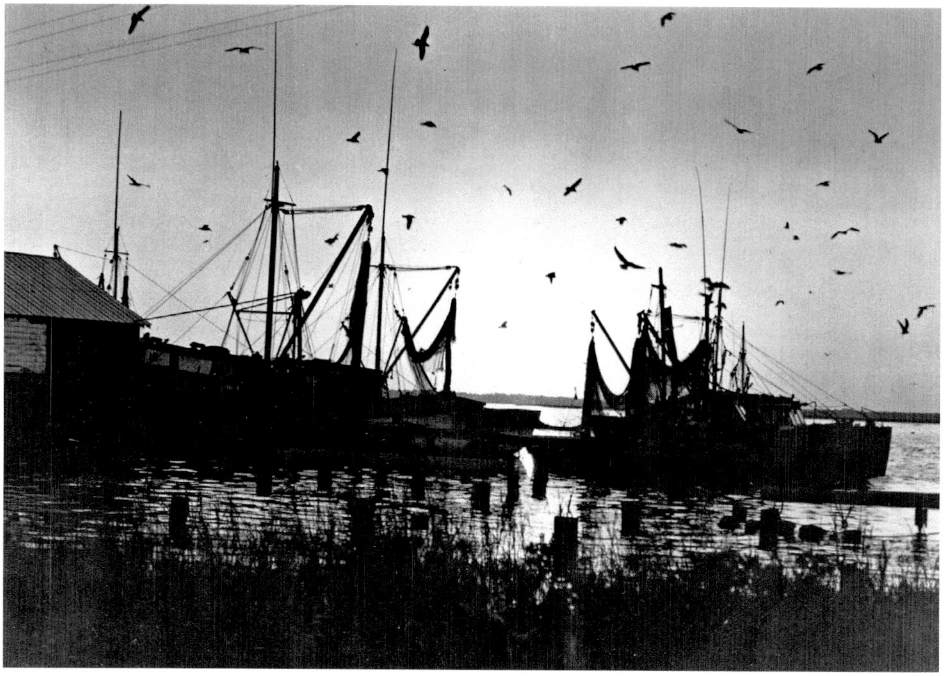

A pair of shrimp boats at dusk on the Amelia River. *AIMH Archives*

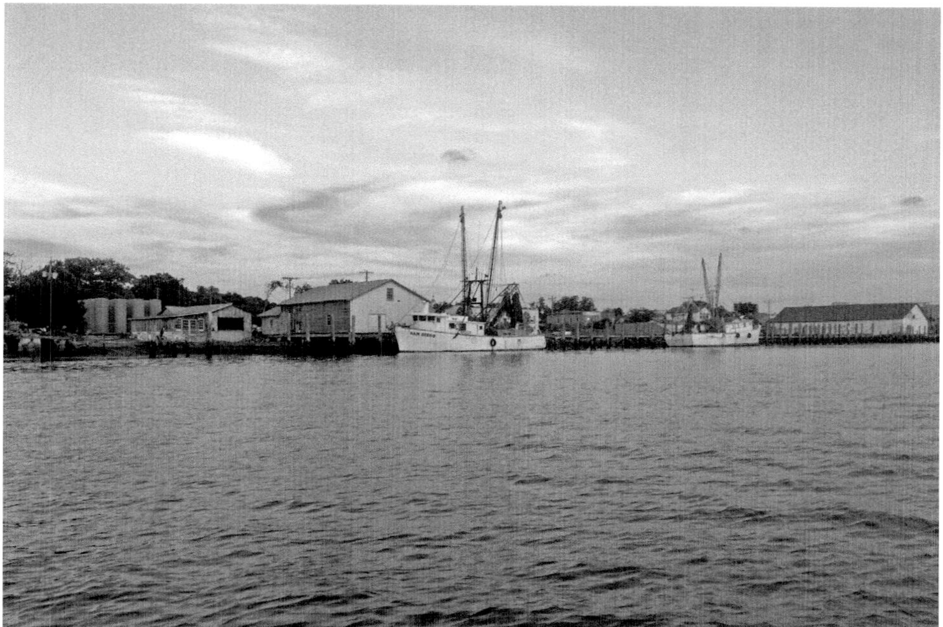

Shrimp Boats docked at Fernandina in 2015. *Photo by Author*

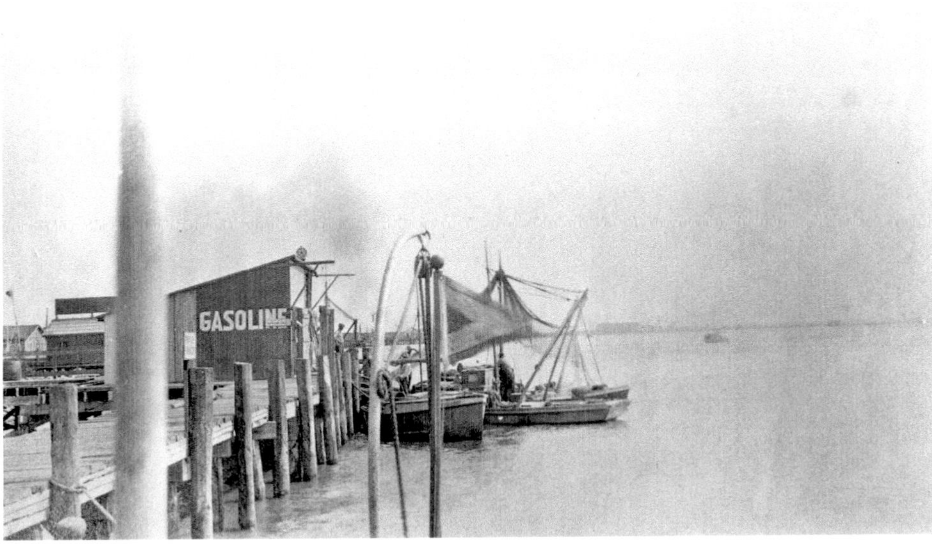

Shrimpers fueling up at Fernandina. *AIMH Archives*

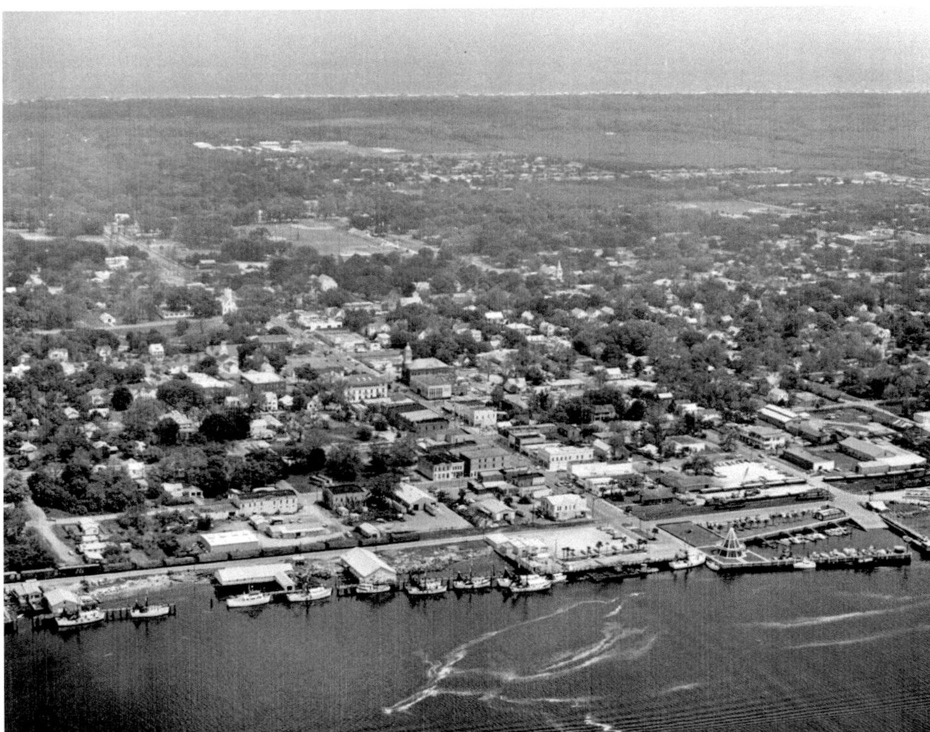

Downtown Fernandina, *c.* 1960s. *Photo by Author*

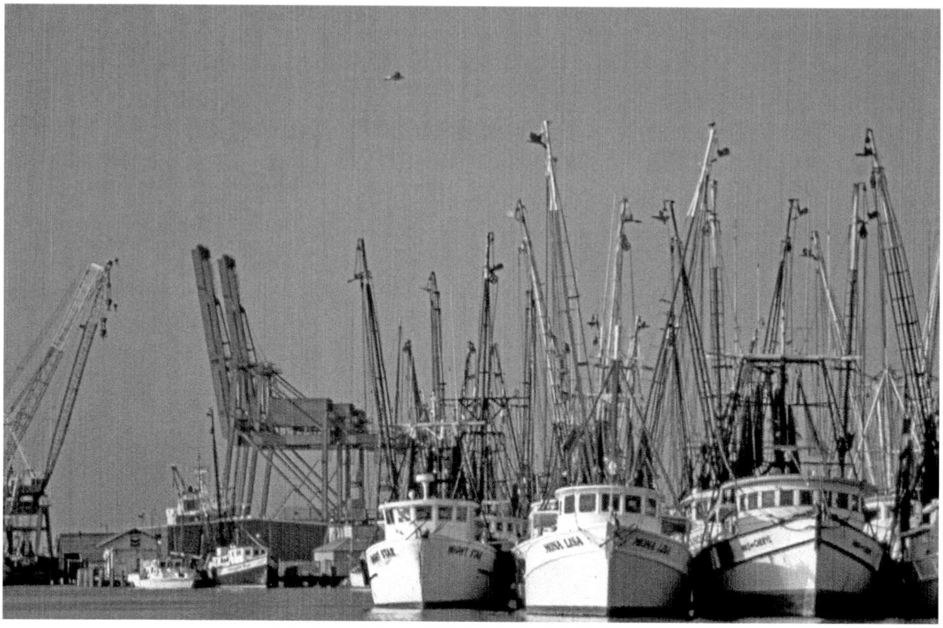

Fleet of Shrimp Boats at Fernandina in 1992. *State Archives of Florida*

Their silhouette became part of the local identity. The drop off in local shrimp boats even since the 1990s is alarming to many, and has forced families that have been shrimping for decades out of the business. By 1995 there were less than twenty shrimp boats left in the area.

The rise of aquaculture has become major competition for wild caught shrimp, as many restaurants and food retailers find it cheaper to import shrimp and other seafood from aqua farms around the world (especially from Asian and South American markets). Americans consume about 1.3 billion pounds of shrimp annually, and only about ten percent (roughly 125 million pounds) of that is wild caught.

Shrimp fisheries are renewed on a yearly basis, but a lean year can prove disastrous for professional shrimpers. Runoff from fertilizers and pesticides can be a detriment to shrimp populations. For two months each year shrimping is open to recreational fishermen. Recreational shrimping can limit the number of shrimp allowed to mature and make their way to the ocean, where most commercial shrimpers trawl.

Despite the difficulties modern commercial shrimpers face, the industry soldiers on. Florida has perennially ranked in top three highest wild caught shrimp producing states until 2009, when Alabama pushed the Sunshine State into fourth place. As of 2014, Florida ranked in fifth place, behind Alabama, Mississippi, Louisiana, and Texas. For many years the Florida's Gulf Coast dominated the east in production, but in 2011 the East Coast was able to edge out the West and account for fifty-one percent of Florida's shrimp. On average, Florida produces around 20,000 pounds of shrimp each year worth roughly 40 million dollars.

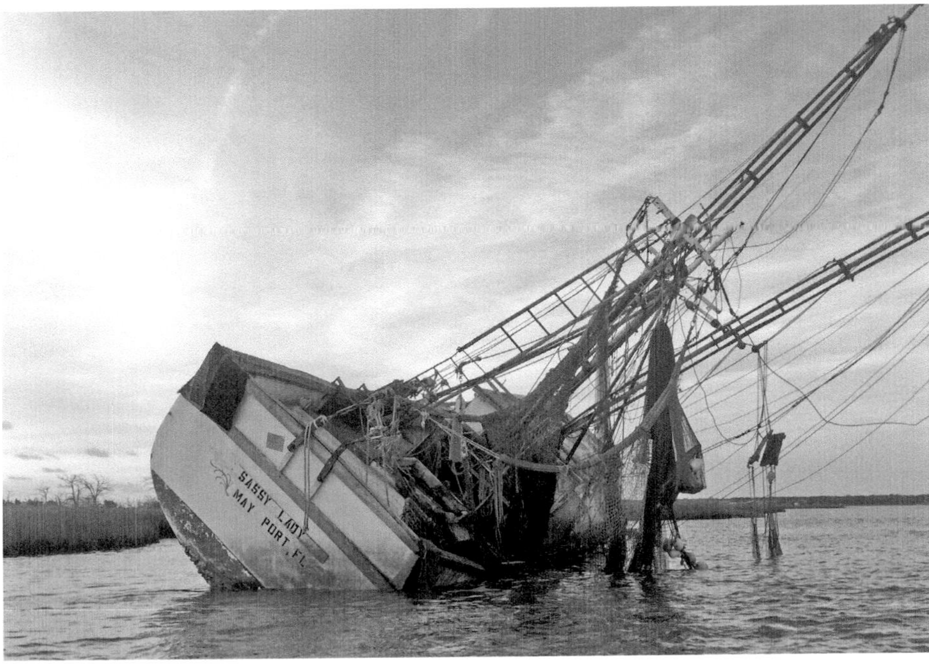

Shrimp boat run aground in the Amelia River, 2015. *Photo by Author*

Many of the pioneering families of Fernandina's shrimping industry – like the Burbanks – have adapted and found ways to survive and succeed in an ever changing industry. The Versaggi Shrimp Corporation is alive and well and operating out of Tampa, Florida. Their six boats produce about 800,000 pounds of shrimp annually, which they sell right off the dock. The Southern Shrimping Association (SSA) was founded in 2002 as an organization of domestic shrimpers, processors and other members of the industry in the eight warm water states who produce shrimp – Florida, Georgia, Alabama, Louisiana, Mississippi, North Carolina, South Carolina, and Texas. Salvatore Versaggi's grandson Sal Versaggi has served as the SSA's president.

There are still local shrimpers fighting to keep the tradition alive. Veteran captain Barry Woods has worked on fishing boats since the age of nine, and though he's followed shrimp runs to Texas, Alabama, Louisiana, and many other places, he still calls Fernandina home. He continues to trawl the Nassau Sound, and his boat is part of the Blessing of the Fleet each year during Shrimp Fest. Second generation shrimper Mike Adams has a dock in nearby Yulee, Florida, and he still shrimps the waterways of Nassau County. In 2012 he took Florida Governor Rick Scott out on the water to demonstrate vital impact commercial shrimping has on Florida's economy.

The Cook family are one of the few that remain from the old guard of local shrimping. They still have docks and trawlers in Fernandina, but they have been acutely affected by

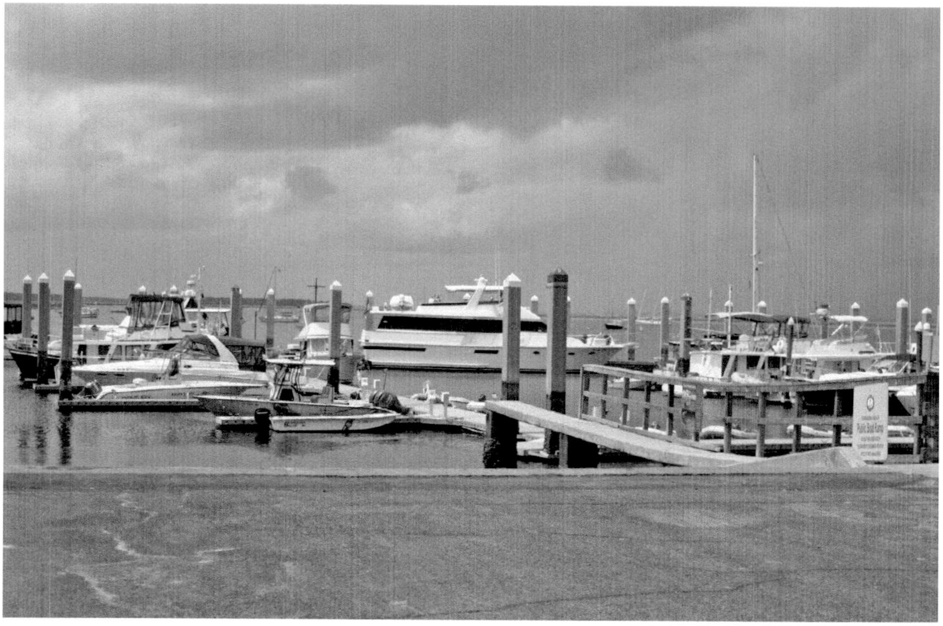

Fernandina's Harbor Marina. *Photo by Author*

the challenges facing the industry. In 2006 Ed Cook (last surviving son of D. A. Cook) then seventy-nine years old, told the *Florida Times-Union* that he and his nephews were looking to retire from shrimping. He summed up the difficulties shrimpers face in the modern era saying: "If you could make a living at shrimping, I'd keep it up for the boys coming up, but with the cost of fuel, the effect that farm-raised shrimp has had on wholesale shrimp prices, and the turtle excluder devices have made it increasingly difficult to earn a living." [10]

Present day, there are around a dozen trawlers left shrimping out of Fernandina on a seasonal basis. Shrimpers caught a break in January of 2005, when the United States International Trade Commission voted to impose tariffs on imported shrimp from China, Brazil, Thailand, Ecuador, and Vietnam. Shrimp coming from these countries (valued at around 2.7 billion dollars annually) is subject to average duties of around seventeen percent.

There are still people in Nassau County who wish to earn a living on the water, like their fathers and grandfathers before them. They hold out hope that new legislation, advances in technology, or natural environmental changes could bring about a resurgence and make local shrimping viable again. When Fernandina's original shrimping interests lulled in 1880, the future of commercial shrimping might have looked just as bleak. The abilities of several determined, hardworking, and progressive thinking individuals have turned the fortunes of Fernandina's shrimping industry around before. As bleak as the future might seem, environmental factors, legislative change, or technological innovations could always light a spark.

Whatever happens to the shrimping industry in the future, its impact on shaping Amelia Island's past is undeniable. In December of 2014, the City of Fernandina Beach partnered with the Amelia Island Museum of History to establish the Fernandina Beach Shrimping Museum and Welcome Center on Front Street. This facility, along with the historic marker that stands just north of the Fernandina Harbor Marina, are manifestations of Amelia Island's reverence for the people from all parts of the shrimping industry who lived and worked here. They are more than just a chapter of local history, they are part of the community's character and identity.

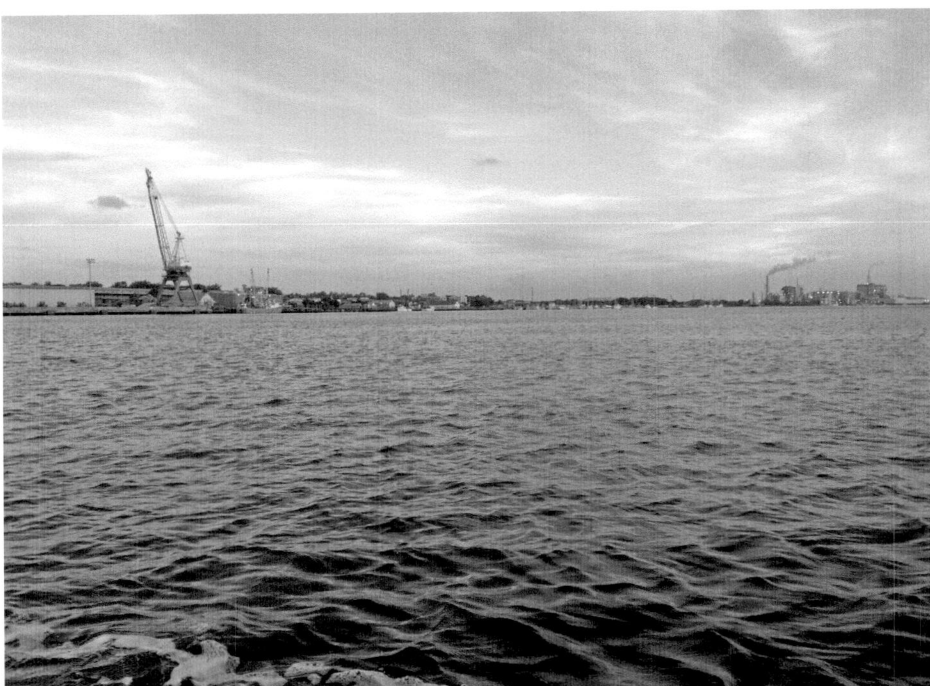

View of Downtown Fernandina taken from the Amelia River in 2015. *Photo Taken by Author*

Acknowledgements

THE HISTORIES PRESENTED HERE belong to all the shrimping families of Fernandina, and I earnestly hope I've done justice to their respective legacies. If this book shines even a glimmer of light on their commendable achievements, the hours of research and writing were time well spent. Without all the people who provided pictures and shared information with the Amelia Island Museum of History Archives over the years, this work would never exist. The descendants of local shrimping pioneers have been phenomenal chroniclers, and they should be commended.

I can't possibly overstate how crucial Nick Deonas and Billy Burbank have been to gathering the information presented here. I marvel at the generosity of these two men in sharing their time, their wealth of knowledge, and their family stories with me. They are not only great ambassadors for this community, but exceptional men. No words of thanks could be enough. I'm proud I grew up in the same town that produced men like these two.

A special thanks to Frank Burbank, for his encyclopedic grasp on fishing and history. The context he provided was indispensable. Thank you Jeff Peterson, for being an expert on all things shrimp and a valuable consultant. Thanks to Susan Hardee Steger for helping me understand the distinguished background of the Hardee family and all their business holdings.

Glenda L. Simons-Jenkins has done a tremendous job researching and keeping alive the legacy of the African American families who helped shape Fernandina's fishing and shrimping history. And she is a gracious steward of that rich tradition. T'ENGKY!

I want to thank the Amelia Island Museum of History Staff: Phyllis Davis, Liz Taylor, Thea Seagraves, Teen Peterson, and Brenda Brubeck, as well as the Museum's Board and volunteers for their patience and for making this possible. No one knows the archives like Teen, and her expertise has saved me hours of research (as have her archival volunteers – you do a tremendous job). You can finally have your space in the archives back.

I want to give a shout out to George Morris at Flying Fish Adventures for getting me in place to take those beautiful pictures from the river. I highly recommend everyone go see Amelia Island via jet-ski with George.

I'd be remiss not to thank my family for their unwavering support. My mom and dad, Helen and Ronnie Edenfield, my sister Aryn DiPietro, and her husband, Joey DiPietro.

My Uncles, Robbie and David Gray, grew up in Fernandina and their experiences helped put life on the island in the 1950s and 1960s into context for me.

No anthology could include the names of all the people who are part of the history of local shrimping. Those who have, so far, gone undocumented or unrecognized deserve to be appreciated for the role they played in developing not only an industry, but a community as well. Even if their names aren't recorded, they will always be part of the story. Everyone who chooses to call Fernandina home owes them a debt of gratitude.

SOURCES

Albert, Charles. Heritage Keepers Project [Audio Recording]. Amelia Island Museum of History Archives. July 24, 2003.

Amelia Now: Winter 1975/76.

Clarke, Gerry. "Shrimping in Fernandina: Yesterday and Today." *Amelia Islander*. June 2010.

Deonas, Nick. *A Pioneer in the Shrimping Industry.* E-book. Fernandina Beach: Search Amelia, 2009.

Daughtry, Angela. "Before the Tide Ebbed – 'Boat Race' Preceded Shrimpfest." *Fernandina News-leader.* May 5, 2013.

Frink, Neil. Heritage Keepers Project [Audio Recording]. Amelia Island Museum of History Archives. September 9, 2003.

Goode, George Brown and a staff of associates. *The Fisheries and Fishing Industries of the United States*. Washington Government Printing Office. 1884.

Hardee, Lewis J. Jr. *Three Southern Families: Hardee-Jones-Davis Families of Coastal North Carolina*. Southport, North Carolina: South Historical Society, 1994.

Hardee, Suzanne Davis (Revised, updated, and edited by Kathleen Davis Hardee Arsenault). *The Golden Age of Amelia Island.* Fernandina Beach, Florida: Lexington Ventures, 2009.

Hurst, Mary. *The Florida Times-Union: Nassau Neighbors Extra.* Thursday, May 11, 2006.

"Fernandina is Birthplace of Modern Shrimping Industry." *Fernandina Beach Fl News-Leader.* November 8, 1995.

Lovejoy, Heather. "Gullah Geechee Commission to Visit Fernandina Thursday." *My Nassau Sun*. February 14, 2009.

Pacetti Jr., Derald. *Shrimping At Fernandina, Florida, before 1920: Industry Development, Fisheries Regulation, Maritime Maturation*. Thesis. August 1980.

Raser, William. "Fernandina's Great Tradition: Down to the Sea in Shrimp Boats." *Amelia Islander Magazine*. May 1997.

Simons-Jenkins, Glenda. "Viewpoint: Gullah/Geechee Culture in Florida." *Fernandina Beach Fl News-Leader.* February 13, 2009.

"Stan Mar – 76 Years of Service to the Shrimp and Fishing Industry." *Fishing Gazette.* February 1976.
"Through the Years with Standard Marine." *The Fish Boat.* January 1977.
Versaggi, John (SDB). *Versaggi Saga.* Manuscript. 1992.
"W. M. Brooks Opens Modern Packing Plant at Fernandina." *The Fish and Oyster Reporter*. October 1936.

ENDNOTES

1 Hardee, Suzanne Davis (Revised, updated, and edited by Kathleen Davis Hardee Arsenault). *The Golden Age of Amelia Island.* Fernandina Beach, Florida: Lexington Ventures, 2009.
2 Pacetti Jr., Derald. *Shrimping At Fernandina, Florida, before 1920: Industry Development, Fisheries Regulation, Maritime Maturation.* Thesis, August 1980.
3 Versaggi, John. *Versaggi Saga.* Manuscript, 1992.
4 Deonas, Nick. *A Pioneer in the Shrimping Industry.* Fernandina Beach: Search Amelia. E-book, 2009.
5 *Biennial Report of the Secretary of State of the State of Florida for the period beginning January 1, 1917, and ending December 31, 1918: Corporations and Drainage Districts Trade Marks.* Tallahassee, Florida: T. J. Appleyard, 1919.
6 "W. M. Brooks Opens Modern Packing Plant at Fernandina." *The Fish and Oyster Reporter*. October 1936.
7 Lovejoy, Heather. "Gullah Geechee Commission to Visit Fernandina Thursday." *My Nassau Sun.* February 14, 2009.
8 Daughtry, Angela. "Before the Tide Ebbed—'Boat Race' Preceded Shrimpfest." *Fernandina Newsleader.* May 5, 2013.
9 Simons-Jenkins, Glenda. "Viewpoint: Gullah/Geechee Culture in Florida." *Fernandina Beach Fl News-Leader.* February 13, 2009.
10 Hurst, Mary. *The Florida Times-Union: Nassau Neighbors Extra.* Thursday, May 11, 2006.